# BASIC PRINCIPLES OF INORGANIC CHEMISTRY

## Making the Connections

# RSC Paperbacks

RSC Paperbacks are a series of inexpensive texts suitable for teachers and students and give a clear, readable introduction to selected topics in chemistry. They should also appeal to the general chemist. For further information on available titles contact

Sales and Promotion Department, The Royal Society of Chemistry, Thomas Graham House, The Science Park, Cambridge CB4 4WF, UK
Telephone: +44 (0)1223 420066
Fax: +44 (0)1223 423623

### New Titles Available

**Archaeological Chemistry**
*by A. M. Pollard and C. Heron*
**Food – The Chemistry of Its Components (Third Edition)**
*by T. P. Coultate*
**The Chemistry of Paper**
*by J. C. Roberts*
**Introduction to Glass Science and Technology**
*by James E. Shelby*
**Food Flavours: Biology and Chemistry**
*by Carolyn L. Fisher and Thomas R. Scott*
**Adhesion Science**
*by J. Comyn*
**The Chemistry of Polymers (Second Edition)**
*by J. W. Nicholson*
**A Working Method Approach for Physical Chemistry Calculations**
*by Brian Murphy, Clair Murphy and Brian J. Hathaway*
**The Chemistry of Explosives**
*by Jacqueline Akhavan*
**Basic Principles of Inorganic Chemistry–Making the Connections**
*by Brian Murphy, Clair Murphy and Brian J. Hathaway*

Existing titles may be obtained from the address below. Future titles may be obtained immediately on publication by placing a standing order for RSC Paperbacks. All orders should be addressed to:

The Royal Society of Chemistry, Turpin Distribution Services Limited, Blackhorse Road, Letchworth, Herts SG6 1HN, UK
Telephone: +44 (0)1462 672555
Fax: +44 (0)1462 480947

RSC Paperbacks

# BASIC PRINCIPLES OF INORGANIC CHEMISTRY
## Making the Connections

BRIAN MURPHY*, CLAIR MURPHY AND
BRIAN J. HATHAWAY

*School of Chemical Sciences, Dublin City University,
Dublin 9, Ireland
and
The Chemistry Department, University College Cork, Ireland

THE ROYAL
SOCIETY OF
CHEMISTRY
Information
Services

ISBN 0–85404–574–0

A catalogue record for this book is available from the British Library

Published by the Royal Society of Chemistry, Thomas Graham House, Science Park, Milton Road, Cambridge CB4 4WF, UK

For further information visit our web site at www.rsc.org
Typeset in Great Britain by Vision Typesetting, Manchester
Printed by Athenaeum Press Ltd, Gateshead, Tyne and Wear, UK

# Preface

With the passage of time, the amount of factual chemistry is continually increasing and it is now virtually impossible for one person to retain. Fortunately, it is not necessary for one person to know all of this, as long as the data can be accessed reasonably quickly. What is more important is that the underlying principles are clearly described and understood. As teachers, it is then the teaching of these principles that should be emphasised in teaching programmes and not the factual data. The latter do have a limited role as reference material, such as in 'The Elements' by J. Emsley, 1989, Oxford University Press and 'The Dictionary of Inorganic Compounds' ed. J. E. Macintyre, 1992, Chapman and Hall, London, but these texts are not appropriate for teaching the basic principles of chemistry. The preparation of textbooks has been made much easier by the improvements in the technology of book production. This has resulted in the production of much more colourfully attractive textbooks, relative to the rather drab texts of 20 years ago, but unfortunately, it has also tended to produce larger textbooks of rarely less than 1000 pages. This is particularly the case with textbooks of general and introductory chemistry. This would not be a problem if the basic principles of chemistry were still clearly identifiable. However, this is rarely the case and the principles, even when well described, are lost beneath a wealth of factually unconnected data that it is of low priority for the student to learn and gives the impression that chemistry is a boring subject. In general, these students are only taking chemistry as a subsidiary subject and will not proceed beyond the basic year.

This is particularly apparent in the sections on introductory inorganic chemistry, where the underlying electron configuration of the elements of the Periodic Table not only determines the Long Form of the Periodic Table, but also determines the physical properties of the elements, atom size, ionisation energies and electron affinities (electron attachment enthalpies), and the chemical properties, characteristic or group oxidation numbers, variable valence and the formation of ionic

v

and covalent bonds. From the valence shell configuration of the Main Group elements in their compounds, the Lewis structure, shape and hybridisation of the bonds in these compounds may be predicted. Equally important, from a knowledge of the valence shell configuration of the elements, the stoichiometry of the reactants and products of the reactions of these elements may be predicted. Such predictions not only form the basic principles for the understanding of preparative inorganic chemistry, they also form the basis of the reactions used in analytical chemistry, namely acid/base, precipitation and redox reactions. Without this understanding of the basic principles of preparative chemistry and chemical reactions, a knowledge of chemistry reduces to pure memory work. Unfortunately, it is this need for pure memory work in learning chemical reactions that forms the basis of teaching in many of the general chemistry textbooks.

The present text tries to overcome the limitations of the above textbooks by covering the basic principles of introductory inorganic chemistry in a structured and connected way, in a short book.

Chapter 1, 'Moles and Molarity', includes a discussion of volumetric calculations, based on *supplied* stoichiometry factors for equations, including limiting reagents. It is included as a first chapter to get students without any previous knowledge of chemistry started on a practical course for volumetric chemistry that usually accompanies an introductory inorganic lecture course.

Chapter 2 describes the 'Structure of the Atom' in terms of electrons and orbitals and the build-up process to the Long Form of the Periodic Table.

Chapter 3 briefly describes how the 'Physical Properties of the Elements' are related to the electron configuration of the elements and hence to their positions in the Periodic Table, namely, their size, ionisation potential and electron attachment enthalpies.

Chapter 4 describes how the 'Chemical Properties of the Elements' are related to their valence shell configuration, *i.e.* characteristic or group oxidation number, variable valence, ionic and covalent bonding. This chapter includes a section on the volumetric calculations used in an introductory inorganic practical course, including the *calculation* of the stoichiometry factors for chemical reactions.

Chapter 5 describes how the Lewis structures of simple Main Group molecules, cations and anions, including oxyacids and oxyanions, are calculated from a knowledge of the valence shell configuration of the central element. A Working Method is suggested for writing the Lewis structures and illustrated by examples, including double bonds and triple bonds in polyatomic molecules.

Chapter 6 describes how the shapes of simple Main Group molecules, cations and anions, including oxyacids and oxyanions, by VSEPR theory are determined from a knowledge of the valence shell configuration of the central element. A Working Method is suggested and illustrated by examples, including double bonds and triple bonds in polyatomic molecules. Given the shapes, hybridisation schemes are suggested to describe the bonding in these covalent species.

Chapter 7 uses the connectivity between the valence shell electron configurations of the elements to systematise the reactivity of the **elements** to form hydrides, oxides and halides, including their molecular stoichiometries. The further reaction of the **compounds** formed is described, using analytical chemistry reactions, *i.e.* acid/base, precipitation and redox reactions, and how the **compounds** behave with water and on heating. A Working Method to describe this *Features of Interest* approach to the descriptive chemistry of molecules is suggested and applied to a number of examples, which are then summarised as **Spider Diagrams**. The use of the Spider Diagram to outline an essay or report on the chemistry of molecules, cations and anions is described.

In University College Cork (UCC), each chapter is accompanied by an interactive computer aided learning tutorial, which briefly introduces each subject, proposes a typical examination question of the appropriate level, and then takes the student stepwise through a suggested Working Method approach to the question, made up of linked multiple-choice questions. Interactive help is provided to each multiple-choice question, with hints provided in the event of an incorrect answer, and up to two attempts are allowed before the correct answer is provided. The Working Method questions are supplemented by independent series of multiple-choice questions. The present short text has been written to discourage the student from using the CAL courseware to take down a set of notes from the computer screen, as this is considered an inappropriate use of these interactive CAL tutorials. These tutorials have been in use for the past four years at University College Cork (and more recently at Cardiff and Dublin City Universities) and are extremely well used by the 300 First Science students per year taking the course. The use of the CAL courseware is entirely optional and supplementary to the normal teaching program, of lectures, practical and large and small group tutorials, but the interactive nature of the courseware, especially for numerical problem solving, is attractive to students, particularly those with a weak chemistry background. As the courseware is based upon UCC type examination questions and also reflects the lecturer's approach to his teaching, the tutorials are not directly transferable to other third-level institutions, but copies are available for down loading from the Internet, free of charge, at:

http://nitec.dcu.ie/~chemlc/CAL2.html

However, these generally follow the approach of the individual chapters in the present text and the authors firmly believe that the **best** courseware should be written in house to reflect the approach of the course lecturer involved.

March 1998          Brian Murphy, Clair Murphy and Brian Hathaway

Brian Murphy          Tel. 353-1-7045472
                      Fax: 353-1-7045503
                      e-mail: murphybr@ccmail.dcu.ie

Clair Murphy          Tel. 353-21-811802
                      Fax: 353-21-811804
                      e-mail: cmurphy@proscom.com

Brian Hathaway        Tel. 353-21-894162
                      Fax: 353-21-270497
                      e-mail: stch8001@bureau.ucc.ie

# Contents

*Chapter 6*
## Shape and Hybridisation    88

*Chapter 7*
## A *Features of Interest* Approach to Systematic
## Inorganic Chemistry    107

# Acknowledgements

The authors wish to acknowledge the help of Professor R.D. Gillard, the Chemistry Department, University of Wales, Cardiff, for comments on an early version of this text, and the continued help and advice of Mrs Janet Freshwater, Books Editor, RSC and Mr A. G. Cubitt, RSC, at the proof stage, However, any remaining errors remain the sole responsibility of the authors.

*Chapter 1*

# Moles and Molarity

## AIMS AND OBJECTIVES

This introductory chapter describes the simple ideas of atoms and molecules, types of chemical formula and their molecular weight for students who have not studied chemistry before. Chemical equations and balanced chemical equations are introduced through the reactions used in an introductory practical laboratory course. The concepts of molarity and molar solutions are introduced through solving volumetric problems, to enable the student to start a laboratory course in practical Inorganic Chemistry.

## STATES OF MATTER

Chemistry is the science and study of the material world. It is generally accepted that there are three states of matter, solid, liquid and gaseous, and the chemicals that make up the materials of the world involve the chemical elements or molecules.

## ELEMENTS, ATOMS AND MOLECULES

The physical state of an element relates to the three states of matter, and the precise state for an element is largely determined by the temperature. Thus at room temperature the element iron is a solid, bromine is a liquid and fluorine is a gas.

In the gaseous state at room temperature helium (He) is a monoatomic gas, and the formula of the element helium is written as He. However, the gaseous form of hydrogen and oxygen at room temperature involves diatomic molecules, namely, $H_2$ and $O_2$. This difference is largely determined by the individual electron configuration of the el-

1

ements, and their ability to form bonds to each other, rather than remain (in the gaseous state) as atomic species of the elements.

The way in which the elements of the **Periodic Table** react together is largely determined by the electron configuration of the individual elements as this determines the ratio in which two elements combine to form a molecule:

$$\text{Atom 1} + \text{Atom 2} \rightarrow \text{Molecule}$$
$$\text{H} + \text{Cl} \rightarrow \text{HCl}$$
$$2\,\text{Atom H} + 1\,\text{Atom O} \rightarrow 1\,\text{Molecule H}_2\text{O}$$

The number of atoms of each element in a molecule determines the ratio of the elements in the molecule and is referred to as the stoichiometry of the molecule. In the molecule of HCl the ratio of H:Cl is 1:1, and the molecule has a stoichiometry of 1:1. In $H_2O$ the ratio of H:O is 2:1, and its stoichiometry is 2:1.

## ELEMENTS, MIXTURES AND COMPOUNDS (MOLECULES)

An **element** consists of only one type of atom, *i.e.* helium, hydrogen or iron. A **mixture** may contain more than one type of substance that can be physically separated into its components, whereas a compound contains more than one type of element, usually with a definite stoichiometry, and cannot be separated into its elements by any simple physical method. Thus the element iron may be obtained as a magnetic black powder that can be mixed with yellow sulfur to give a blackish yellow mixture, from which the iron metal can be separated by means of a magnet. However, if the mixture is heated, a reaction occurs to give a black solid of FeS, iron(II) sulfide, on cooling, from which the iron present cannot be separated by the use of a magnet. The black solid FeS is referred to as a **compound** of Fe and S which has lost the properties of the elemental Fe and S and has unique properties of its own. Similarly, molecules of $H_2$ and $O_2$ react to give molecules of water, $H_2O$:

$$2H_{2(g)} + O_{2(g)} \rightarrow 2H_2O_{(1)}$$

but while $H_2$ and $O_2$ are gases at room temperature, $H_2O$ is a liquid. In these new **compounds** the compound elements are said to have reacted chemically together to give a new compound, FeS and $H_2O$, respectively, with definite stoichiometries between the atoms, namely, 1:1 in FeS and 2:1 in $H_2O$.

## SIMPLE CHEMICAL NAMES

The most simple compounds are those which contain only two elements, one metallic and one non-metallic (explained later). The metal is given the full element name, and the non-metal has the ending -ide.

Thus:

|       |                  |
|-------|------------------|
| NaCl  | sodium chloride  |
| MgO   | magnesium oxide  |
| CaS   | calcium sulfide  |
| BN    | boron nitride    |

If the stoichiometry of the two elements is not $1:1$, prefixes are used thus:

| $1:1$ | mono  | – carbon monoxide           | $CO$     |
|-------|-------|-----------------------------|----------|
| $1:2$ | di    | – carbon dioxide            | $CO_2$   |
| $1:3$ | tri   | – sulfur trioxide           | $SO_3$   |
| $1:4$ | tetra | – carbon tetrachloride      | $CCl_4$  |
| $1:5$ | penta | – phosphorus pentachloride  | $PCl_5$  |
| $1:6$ | hexa  | – sulfur hexafluoride       | $SF_6$   |

**Note:** where more than one atom is present the number is written as a post-subscript.

Compounds with more than two elements cannot end in -ide and for those where the third element is oxygen, the endings -ite or -ate are used:

|                    |          |
|--------------------|----------|
| magnesium sulfide  | $MgS$    |
| magnesium sulfite  | $MgSO_3$ |
| magnesium sulfate  | $MgSO_4$ |

## CATIONS AND ANIONS

In compounds such as NaCl, the lattice is made up of cations (positively charged species) of $Na^+$ and anions (negatively charged species) of $Cl^-$, $Na^+Cl^-$, such that the formula, NaCl, has an overall neutral charge. In $Na_2SO_4$ the overall neutral charge is maintained, but the compound contains two $Na^+$ cations to one $SO_4^{2-}$ anion, with the latter referred to as an oxyanion, in this case a sulfate oxyanion. In aqueous solution the oxyanions occur as discrete species, in the case of the sulfate anion with a $2-$ negative overall charge.

## TYPES OF CHEMICAL FORMULA

In chemistry, different types of chemical formula are used to give different types of information.

   (a) **Empirical Formula:** this is the simplest whole number ratio of the atoms in a molecule; thus in ethanoic acid the empirical formula is $CH_2O$.
   (b) **Molecular Formula:** this is the actual number of atoms making up the molecule; thus in ethanoic acid the molecular formula is $C_2H_4O_2$, *i.e.* twice the empirical formula.
   (c) **Structural Formula:** this shows the various ways of representing the actual arrangement of atoms in the molecule, *i.e.*

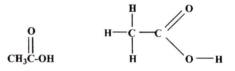

CH₃COOH                    CH₃C-OH

## ATOMIC WEIGHT

The atomic weight or relative atomic mass of an element is the mass of one atom of that element relative to that of the most abundant form of carbon taken as 12 units. On this scale the atomic weight of hydrogen is 1, oxygen is 16, and copper is 63.54 a.m.u. Table 1.1 lists the atomic weights of the first 18 elements of the Periodic Table.

On this scale the molecular formula of ethanoic acid, $C_2H_4O_2$ has a molecular weight of:

$$2C(12) + 4H(1) + 2O(16), i.e. (24 + 4 + 32 = 60)$$

namely, 60 atomic mass units (a.m.u.), *i.e.* the gram mole or molecular weight of ethanoic acid is 60. The gram mole of ethanoic acid is then 60 g and contains:

24 gram atoms of carbon
4 gram atoms of hydrogen
32 gram atoms of oxygen
Total: 60 grams.

**Table 1.1** *The atomic weights of the first 18 elements of the Periodic Table*

| Hydrogen | 1.008 | Helium | 4.003 | Lithium | 6.941 |
|---|---|---|---|---|---|
| Beryllium | 9.012 | Boron | 10.811 | Carbon | 12.011 |
| Nitrogen | 14.007 | Oxygen | 16.000 | Fluorine | 18.998 |
| Neon | 20.179 | Sodium | 22.990 | Magnesium | 24.305 |
| Aluminium | 26.982 | Silicon | 28.086 | Phosphorus | 30.974 |
| Sulfur | 32.066 | Chlorine | 35.453 | Argon | 39.948 |

## AVOGADRO'S NUMBER

As the gram mole of a molecule (60 for ethanoic acid) is defined **relative** to the gram atom of carbon $= 12\,g$, the actual number of atoms in $12\,g$ carbon has been experimentally determined as $6.022 \times 10^{23}$ atoms. This is referred to as **Avogadro's Number**, and is the number of atoms in the gram atomic weight of any element, *i.e.* 19 g fluorine, 32 g sulfur or 63.5 g copper. It then follows that the number of molecules in the gram molecular weight of a molecule (1 gram mole) is also $6.022 \times 10^{23}$, Avogadro's Number. Thus one mole of ethanoic acid, 60 g, contains $6.022 \times 10^{23}$ molecules of ethanoic acid. Equally, one mole of dihydrogen, $H_2$, 2 g, one mole of water, $H_2O$, 18 g, and one mole of sulfuric acid $H_2SO_4$, 98 g, each contains $6.022 \times 10^{23}$ molecules.

It also follows that 1 g of a molecule will contain Avogadro's Number divided by the gram molecule weight (1 mole) of the molecule:

$$\therefore \text{ 1 g ethanoic acid contains } 6.022 \times 10^{23}/60$$
$$\text{molecules} = 1.0037 \times 10^{22} \text{ molecules}$$

Likewise:

$$\text{1 g hydrogen (0.5 1 gram mole) contains } 3.011 \times 10^{23} \text{ molecules}$$
$$\text{1 g sulfuric acid (1/98 1 gram mole) contains } 6.145 \times 10^{21} \text{ molecules.}$$

## EMPIRICAL FORMULA

This only expresses the relative number of atoms of each element in a compound. Nevertheless, it is the first step in the experimental determination of the molecular formula of a compound from its percentage composition.

1. Thus: A contains 42.9% C and 57.1% O; calculate its empirical/ molecular formula

|          | Atomic Wt. | %    | %/At. Wt.         | Ratio |
|----------|------------|------|-------------------|-------|
| Carbon   | 12         | 42.9 | 42.9/12 = 3.58    | 1     |
| Oxygen   | 16         | 57.1 | 57.1/16 = 3.58    | 1     |

$\therefore$ Empirical formula is $C_1O_1$ or CO (carbon monoxide).

2. A contains 79.9% C and 20.1% H:

|          | Atomic Wt. | %    | %/At. Wt.          | Ratio |        |
|----------|------------|------|--------------------|-------|--------|
| Carbon   | 12         | 79.9 | 79.9/12 = 6.67     | 1     |        |
| Hydrogen | 1          | 20.1 | 20.1/1 = 20.1      | 3     | $CH_3$ |

3. A contains 37.5% C; 12.5% H; 50.0% O:

|          | Atomic Wt. | %    | %/At. Wt.          | Ratio |          |
|----------|------------|------|--------------------|-------|----------|
| Carbon   | 12         | 37.5 | 37.5/12 = 3.12     | 1     |          |
| Hydrogen | 1          | 12.5 | 12.5/1 = 12.5      | 4     |          |
| Oxygen   | 16         | 50.0 | 50.0/16 = 3.12     | 1     | $CH_4O$  |

4. A contains 43.7% P; 56.3% O:

|            | Atomic Wt. | %    | %/At. Wt.       | Ratio |           |
|------------|------------|------|-----------------|-------|-----------|
| Phosphorus | 31         | 43.7 | 43.7/31 = 1.4   | 2     |           |
| Oxygen     | 16         | 56.3 | 56.3/16 = 3.5   | 5     | $P_2O_5$  |

5. Given the molecular formula of ethanoic acid, $CH_3CO_2H$ what percentages of C, H and N are present?

$$CH_3CO_2H \equiv C_2H_4O_2 \equiv 2 \times CH_2O$$

Empirical weight $= 12 + 2 + 16 = 30$ and the molecular weight $= 24 + 4 + 32 = 60$.

% C $= 24/60 \times 100 = 40.0\%$ C
% H $= 4/60 \times 100 = 6.67\%$ H
% O $= 32/60 \times 100 = 53.3\%$ O

## CHEMICAL EQUATIONS

Chemistry involves the study of the ways in which the elements and compounds react with each other. We have already seen:

Fe          + S          → FeS
1 atom      1 atom       1 molecule
$2H_2$      + $O_2$      → $2H_2O$
2 molecules  1 molecule  2 molecules

in which two pairs of elements react to form a compound. Some more complicated balanced equations are:

$$Na_2SO_4 + BaCl_2 → BaSO_4 + 2NaCl$$
1 mole      1 mole   1 mole    2 moles

Notice because of the balancing of charges, 1 mole of each of the reactants produces 2 moles of NaCl. Equally:

2Al        + 6HCl         → $2AlCl_3$      + 3 $H_2$
2 atoms     6 molecules     2 molecules     3 molecules

Such reactions contain a great deal of information; thus in the reaction:

$N_2$        + $3H_2$         → $2NH_3$
1 molecule    3 molecules     2 molecules

could be represented alternatively:

in a structural notation. It also contains more quantitative information:

1. 1 mole $N_2$ reacts with 3 moles $H_2$ to give 2 moles $NH_3$;
2. 28 g (1 mole) $N_2$ reacts with 6 g (3 moles) $H_2$ to give 34 g (2 moles) $NH_3$;
3. 1 g $N_2$ requires 6/28 g $H_2$ for complete reaction to give 34/28 g $NH_3$;
4. 1 g $N_2$ in excess $H_2$ will only yield 34/28 g $NH_3$.

## BALANCING CHEMICAL EQUATIONS

Such chemical equations must obey certain rules:

1. The reactants are written to the left-hand side, LHS, the products

to the right-hand side, RHS, of the reaction arrow →.
2. Each side of the equation must have the same number of each kind of atoms, *i.e.* the equation must balance.
3. The common gaseous elements are shown as diatomic – $H_2$, $O_2$, $N_2$, $Cl_2$ – and solid elements as atoms – C, P, S, Cu or alternatively as $C_x$, $P_4$, $S_8$, $Cu_x$.
4. The overall ionic charges must be the same on each side of the equation.

For example, to balance the equation:

$$Al + HCl \rightarrow AlCl_3 + H_2.$$

steps 1–4 must be followed:

1. The products involve 3Cl, while the reactants involve only 1Cl
   ∴ $Al + 3HCl \rightarrow AlCl_3 + H_2$
2. The reactants involve 3H, the products 2H
   ∴ $Al + 2 \times 3HCl \rightarrow AlCl_3 + 3H_2$
3. The reactants involve 6Cl, the products 3Cl
   ∴ $2Al + 6HCl \rightarrow 2AlCl_3 + 3H_2$

and the equation is now balanced.

## MOLAR SOLUTIONS

One of the values of the term mole is that it can be used as a measure of concentration in solution. Namely, 1 gram mole of a molecule dissolved in $1000 \, cm^3$ is defined as a 1 molar solution 1 M, thus:

$$60 \, g \text{ ethanoic acid in } 1000 \, cm^3 \equiv 1 \, M$$
$$30 \, g \text{ ethanoic acid in } 1000 \, cm^3 \equiv 0.5 \, M$$
$$15 \, g \text{ ethanoic acid in } 1000 \, cm^3 \equiv 0.25 \, M$$

Equally:

$1000 \, cm^3$ of a 1 M solution of ethanoic acid contains 60 g ethanoic acid (1 mole);

$500 \, cm^3$ of a 1 M solution of ethanoic acid contains 30 g ethanoic acid (0.5 mole);

$250 \, cm^3$ of a 1 M solution of ethanoic acid contains 15 g ethanoic acid (0.25 mole).

In general the number of moles of a substance A in $V_A$ cm$^3$ of a $M_A$ molar solution is given as:

$$\text{Number of moles of A} = V_A M_A/1000$$

as 1000 cm$^3$ of a 1 molar solution contains one gram mole.

From the definition of a molar solution, *i.e.* the number of moles per litre, the number of moles of a reagent is related through the molecular weight to the weight of the reagent present in a known volume, $V_A$, and molarity, $M_A$, by the relationship:

$$V_A \times M_A/1000 \times \text{molecular weight of A} = \text{weight of A in } V_A \text{ cm}^3 \text{ of solution.}$$

Thus the weight of $H_2SO_4$ in 23.78 cm$^3$ of a 0.123 M solution of sulfuric acid, molecular weight 98.070, will be:

$$23.78 \times 0.123/1000 \times 98.070 = 0.2869 \text{ g.}$$

## VOLUMETRIC REACTIONS

In the laboratory three general types of titration reactions are met with, namely:

(a) acid/base reactions:

$$HCl + NaOH \rightarrow NaCl + H_2O$$
$$H_2SO_4 + 2NaOH \rightarrow Na_2SO_4 + 2H_2O$$
$$H_3PO_4 + 3NaOH \rightarrow Na_3PO_4 + 3H_2O$$
$$2HCl + Ca(OH)_2 \rightarrow CaCl_2 + 2H_2O$$
$$2H_3PO_4 + 3Ca(OH)_2 \rightarrow Ca_3(PO_4)_2 + 6H_2O$$

(b) precipitation:

$$AgNO_3 + NaCl \rightarrow AgCl \downarrow$$
$$Ba(OH)_2 + H_2SO_4 \rightarrow BaSO_4 \downarrow$$
$$3Ag_2SO_4 + 2AlCl_3 \rightarrow 6AgCl \downarrow$$

(c) redox reactions:

$$KMnO_4 + 5FeSO_4 \rightarrow Mn^{2+} + 5Fe^{3+}$$
$$K_2Cr_2O_7 + 6FeSO_4 \rightarrow 2Cr^{3+} + 6Fe^{3+}$$

All of these reactions may be expressed in the general form:

$$v_A \cdot A + v_B \cdot B \rightarrow$$

with the number of moles of the two reactants given separately as:

$$\frac{V_A \cdot M_A}{1000} \text{ and } \frac{V_B \cdot M_B}{1000}$$

However, unless the stoichiometry factors $v_A$ and $v_B$, respectively, are identically both equal to one, these numbers of moles are NOT identical. They can be equated if the expressions are modified by the respective stoichiometry factors $v_A$ and $v_B$ as:

$$\frac{V_A \cdot M_A}{1000} \cdot \frac{1}{v_A} = \frac{V_B \cdot M_B}{1000} \cdot \frac{1}{v_B}$$

## VOLUMETRIC TITRATIONS

The relationship:

$$\frac{V_A \cdot M_A}{1000} \cdot \frac{1}{v_A} = \frac{V_B \cdot M_B}{1000} \cdot \frac{1}{v_B} \tag{1}$$

may be used to calculate an unknown quantity for the reaction:

$$v_A \cdot A + v_B \cdot B \rightarrow v_C \cdot C + v_D \cdot D$$

when a volume $V_A$ of a solution of molarity $M_A$ is titrated by a volume $V_B$ of a solution of molarity $M_B$. If three of the four unknowns, $V_A$, $M_A$, $V_B$, and $M_B$ are provided, the value of the fourth can be calculated, **provided** the values of $v_A$ and $v_B$ are known.

Thus, in the question: When $25\,cm^3$ of a 0.176 molar solution of $H_3PO_4$ is titrated by a 0.123 molar solution of $Ca(OH)_2$, what volume of the latter is required?

Given the reaction:

$$2H_3PO_4 + 3Ca(OH)_2 \rightarrow$$

then:

$$\frac{25.00 \times 0.176}{1000} \cdot \frac{1}{2} = \frac{V_D \times 0.123}{1000} \cdot \frac{1}{3}$$

$$V_B = \frac{25.00 \times 0.176 \times 3}{0.123 \times 2} = 53.66 \, \text{cm}^3$$

Equation (1) may then be used to calculate any unknown out of the four variables, $V_A$, $M_A$, $V_B$ and $M_B$, but to do so it is essential that the stoichiometry factors $v_V$ and $v_B$ are known.

**Note:** given $v_A$ and $v_B$ it is unnecessary to know $v_C$ and $v_D$, but these can be evaluated.

## LIMITING REAGENTS

Under certain conditions, all of the reagents in a chemical reaction may not be completely consumed.

$$2AgNO_3 + Cu \rightarrow 2Ag + Cu(NO_3)_2$$

The equation indicates that 2 moles of $AgNO_3$ must react with 1 mole of Cu to give the products. If there were 2 moles of $AgNO_3$ and 2 moles of Cu, then 1 mole of Cu must remain unreacted at the end of the reaction since 2 moles of $AgNO_3$ can only react with 1 mole of Cu.

$$2AgNO_3 + 2Cu \rightarrow 2Ag + Cu(NO_3)_2 + \textbf{Cu} \tag{1}$$

If 3 moles of $AgNO_3$ and 1 mole of Cu are reacted, then 1 mole of $AgNO_3$ must remain unreacted at the end of the reaction.

$$3AgNO_3 + Cu \rightarrow 2Ag + Cu(NO_3)_2 + \textbf{AgNO}_3 \tag{2}$$

since only 2 moles of $AgNO_3$ can react with one mole of Cu.

The reactant that is completely consumed in the reaction is termed the **limiting reagent**, *i.e.* the $AgNO_3$ limits the amount of product that can be formed in reaction (1) and the Cu in reaction (2). The other reactants are present in **excess**.

## Worked Example No. 1

If a mixture of 10.0 g of Al and 50.0 g of $Fe_2O_3$ react with each other to produce $Al_2O_3$ and Fe, how many grams of iron are produced?

**Solution:**

(i) Balanced Equation:

$$2Al + Fe_2O_3 \rightarrow 2Fe + Al_2O_3$$

(ii) From the Balanced Equation:

2 moles of Al reacts completely with 1 mole of $Fe_2O_3$
to produce 2 moles of Fe
*i.e.* 2 moles Al $\equiv$ 1 mole $Fe_2O_3$ $\equiv$ 2 moles Fe $\equiv$ 1 mole $Al_2O_3$.

(iii) How many moles of Al are present?

1 mole of Al = Atomic Weight of Al
$\therefore$ 1 mole of Al = 27 g of Al
$\rightarrow$ 10 g of Al = 10/27 moles of Al = 0.37 moles

Answer: There are 0.37 moles of Al present.

(iv) How many moles of $Fe_2O_3$ are required by 0.37 moles Al?

2 moles Al $\equiv$ 1 mole $Fe_2O_3$
$\rightarrow$ 0.37 moles Al $\equiv$ 0.185 moles $Fe_2O_3$

Answer: 0.185 moles of $Fe_2O_3$ are needed.

(v) How many moles of $Fe_2O_3$ are present?

1 mole = Molecular Weight
$\therefore$ 1 mole = 159.6 g
$\rightarrow$ 50 g of $Fe_2O_3$ = 50/159.6 moles of $Fe_2O_3$ = 0.313 moles

Answer: 0.313 moles of $Fe_2O_3$ are present.

$\therefore$ $Fe_2O_3$ is present in excess.

(vi) How many moles of Al are required when 0.313 moles $Fe_2O_3$ are present?

1 mole $Fe_2O_3$ $\equiv$ 2 moles Al
0.313 moles $Fe_2O_3$ $\equiv$ 0.626 moles Al

Answer: 0.626 moles of Al are required, but as there are only 0.37 moles of Al present; therefore Al is the limiting reagent.

(vii) Since Al is the limiting reagent and only 0.37 moles are present, then from the balanced equation:

$2Al \rightarrow 2Fe$
0.37 moles of Al $\equiv 2 \times 0.37/2$ moles of Fe $= 0.37$ moles of Fe

Answer: Only 0.37 moles of Fe can be produced from a mixture of 10 g of Al and 50 g of $Fe_2O_3$.

**Note:** To determine which reagent is the limiting reagent, calculate the amount of product expected from each reactant. The reactant that gives the smallest amount of product is the limiting reagent.

### Worked Example No. 2

15 g of a substance P, 23 g of a substance Q and 10 g of a substance R react together, completely, to form a product S. How much S will be produced from a mixture of 0.049 g of P, 0.029 g of Q and 0.37 g of R?

**Solution:**

(i) Equation:

$15g\,P + 23g\,Q + 10g\,R \rightarrow$

Since the reactants react **completely** to form S, there is

$15g + 23g + 10g$ of S formed $= 48g$ of S.

(ii) To find the limiting reagent:

$$15g\,P + 23g\,Q + 10g\,R \rightarrow 48g\,S$$
$$0.04g\,P \rightarrow 48/15 \times 0.04 = 0.128\,g\,S$$
$$0.029\,Q \rightarrow 23/15 \times 0.029 = 0.042\,g\,S$$
$$0.37\,R \rightarrow 10/15 \times 0.37 = 0.78\,g\,S$$

Answer: The limiting reagent is Q, since the amount of Q present only produces 0.042 g S and some of the reagents P and R will remain in excess at the end of the reaction.

*Chapter 2*

# The Structure of the Atom, Electron Configuration and the Build-up to the Periodic Table

## AIMS AND OBJECTIVES

This chapter introduces the electronic structure of the atom, from the early shell structure of the Bohr theory, using the single principal quantum $n$, through the wave nature of the electron, the Schrödinger wave equation, and the need for the four quantum numbers, $n$, $l$, $m_l$ and $m_s$, to describe the occurrence of the $s$, $p$, $d$ and $f$ orbitals. The evidence for this more complicated shell structure is seen in the photoelectronic spectra of the elements; this justifies the one electron orbital description of the atom and from which the $s$-, $p$-, $d$- and $f$- block structure of the Periodic Table is developed.

## THE STRUCTURE OF THE ATOM

The material world is made up of atoms, molecules and ions. The first reference to atoms can be found in the writings of the ancient Greeks. The first clear atomic hypothesis for the existence of atoms, was presented in 1805 by John Dalton. He suggested:

1. All atoms of a given element are identical;
2. The atoms of different elements have different masses;
3. A compound is a specific combination of atoms of more than one element;
4. In a chemical reaction atoms are neither created nor destroyed, but merely exchange partners.

Dalton's 'hypothesis' was a suggestion to account for the observed

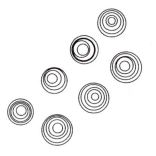

**Figure 2.1** *Atoms can be seen as bumps on the surface of a solid using the electron tunnelling microscope*

**Table 2.1** *The fundamental atomic particles*

|  | *Mass*/a.m.u. | *Charge* |
|---|---|---|
| Electron | 0.00055 | − 1 |
| Proton | 1.0073 | + 1 |
| Neutron | 1.0087 | 0 |

combining masses of the elements that formed compounds. To understand this it is necessary to know about the structure of the atom.

The evidence for the atom is now direct, as it is possible to see atoms directly, using such techniques as electron tunnelling microscopy. If this technique is used to look at the surface of copper metal, the atoms show up as bumps (Figure 2.1). The atom may be defined as the smallest unit of an element that retains the physical and chemical characteristics of the element. Dalton considered that the atom could be treated as a hard sphere that could not be broken down into smaller units, *i.e.* it had no internal structure, rather like a billiard ball. While this is not quite true, it can be understood in terms of the present knowledge of the structure of the atom. In the late 1800s, J. J. Thompson showed that the atom was built up from much smaller units, namely, electrons, protons and neutrons (Table 2.1).

The electron carries a single negative charge, the proton a single positive charge and the neutron no charge. The proton and neutron have a comparable mass of 1 atomic mass unit (a.m.u.), approximately that of the hydrogen atom. The electron is much smaller, 0.0055 a.m.u. (1/1837 a.m.u.). The unit of mass of the atom is called the **atomic mass unit** (a.m.u.) and is taken to be exactly 1/12th of the carbon-12 isotope (see later). Hence, the mass of carbon-12 is 12 a.m.u., or $1.9927 \times 10^{-23}$ grams. 1 a.m.u. is equal to $1.6606 \times 10^{-24}$ grams.

J. J. Thompson initially visualised the atom as electrons embedded in

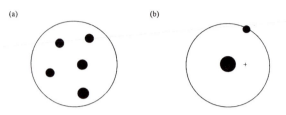

**Figure 2.2** *The structure of the atom:* (a) *J. J. Thomson's 'Plum Pudding' model;* (b) *N. Bohr's model*

**Table 2.2** *Some elements and isotopes*

| Element | Z | Protons | Electrons | Neutrons | Mass | Symbol |
|---|---|---|---|---|---|---|
| Hydrogen | 1 | 1 | 1 | 0 | 1 | H |
| Deuterium | 1 | 1 | 1 | 1 | 2 | H(D) |
| Tritium | 1 | 1 | 1 | 2 | 3 | H(T) |
| Helium | 2 | 2 | 2 | 2 | 4 | He |
| Lithium | 3 | 3 | 3 | 4 | 7 | Li |
| Beryllium | 4 | 4 | 4 | 5 | 9 | Be |

a sea of positively charged jelly, *i.e.* the 'plum pudding' model of the atom (Figure 2.2a). However, this was soon changed to the present day view of the atom as involving a small positively charged nucleus consisting of protons and neutrons, surrounded by negatively charged electrons (Figure 2.2b). The mass of an atom is largely concentrated in the central nucleus made up of protons (positively charged), and neutrons (no charge), and surrounded by electrons (negatively charged). Hence, the nucleus carries a positive charge, which attracts the outer electron, and the positive charge is balanced by the appropriate number of electrons (negatively charged). The number of protons in the nucleus determines the atomic number, $Z$, of an element, and the number of protons is approximately equal to the number of neutrons. Hence the approximate mass of an element is $\sim 2Z$, and is largely concentrated in the nucleus. The atomic number $Z$ determines the type of element involved, some examples of which are shown in Table 2.2.

Particles with the same atomic number $Z$, but different numbers of neutrons are called **isotopes**. Deuterium and tritium are **isotopes** of hydrogen, as they all have the same atomic number, $Z = 1$, but different numbers of neutrons, namely, 0, 1 and 2, respectively, and have different mass numbers of 1, 2 and 3, respectively. While the mass numbers are integer, the atomic weights are not necessarily integer if more than one isotope of an element occurs naturally. Thus, Cl (atomic weight = 35.46)

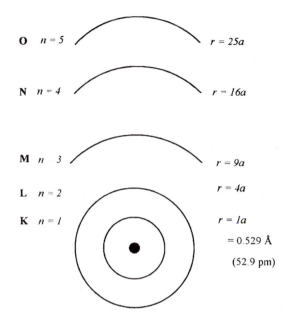

O   $n = 5$     $r = 25a$

N   $n = 4$     $r = 16a$

M   $n$   3     $r = 9a$

L   $n = 2$     $r = 4a$

K   $n = 1$     $r = 1a$

$= 0.529$ Å

(52.9 pm)

**Figure 2.3** *A sketch of the circular orbits of the Bohr model of the hydrogen atom*

is made up of the sum of 77.00% $^{35}_{17}Cl$ and 23.00% $^{37}_{17}Cl$, where 35 and 37 refer to the mass numbers of the pure isotopes and 17 refers to the atomic number. Natural magnesium consists of 79% $^{24}_{12}Mg$, 10% $^{25}_{12}Mg$, and the atomic weight of Mg = 24.32, rather than exactly 24.00.

## BOHR MODEL OF THE ATOM

The chemical properties of the atom are determined by the number of valence shell electrons ($Z$) in an atom, and the way these electrons are arranged in electron shells. The simple Bohr theory quantised the energies of the electrons into discrete **K, L, M, N,** and **O** shells (Figure 2.3). This shell theory also allows the prediction of the number of electrons per shell as $2n^2$ electrons, namely, 2, 8, 18, 32, *etc.* electrons, respectively, as shown in Table 2.3, where $n$ is now referred to as the *principal quantum number*.

Table 2.4 shows some examples of the alternative spherical shell description of these electronic configurations.

Some spectroscopic evidence for this shell structure of the valence electrons of the elements can be seen in the line structure, Figure 2.4, of

**Table 2.3** *The occupation of electron shells* – $2n^2$

| Shell | K | L | M | N | O |
|-------|---|---|---|---|---|
| $n$ | 1 | 2 | 3 | 4 | 5 |
| $2n^2$ | 2 | 8 | 18 | 32 | 50 |

**Table 2.4** *The electron configurations of the elements in the first four shells, using the K, L, M and N shell configurations*

| Element | Z | Electron shells | | | |
|---------|---|---|---|---|---|
| | | K | L | M | N |
| $n$ | | 1 | 2 | 3 | 4 |
| Hydrogoen | 1 | 1 | | | |
| Helium | 2 | 2 | | | |
| Lithium | 3 | 2 | 1 | | |
| Beryllium | 4 | 2 | 2 | | |
| Boron | 5 | 2 | 3 | | |
| Carbon | 6 | 2 | 4 | | |
| Nitrogen | 7 | 2 | 5 | | |
| Oxygen | 8 | 2 | 6 | | |
| Fluorine | 9 | 2 | 7 | | |
| Neon | 10 | 2 | 8 | | |
| Sodium | 11 | 2 | 8 | 1 | |
| Magnesium | 12 | 2 | 8 | 2 | |
| Aluminium | 13 | 2 | 8 | 3 | |
| Silicon | 14 | 2 | 8 | 4 | |
| Phosphorus | 15 | 2 | 8 | 5 | |
| Sulfur | 16 | 2 | 8 | 6 | |
| Chlorine | 17 | 2 | 8 | 7 | |
| Argon | 18 | 2 | 8 | 8 | |
| Potassium | 19 | 2 | 8 | 8 | 1 |

the hydrogen atom, where the energies are given by $E = -kZ^2/n^2$ and the differences in energies of the observed spectra by expressions:

$$E_2 - E_1 = h\nu = hZ^2(1/n_1{}^2 - 1/n_2{}^2)$$

where the integers refer to $n$, the principal quantum numbers associated with a particular shell, and $\nu$ refers to the frequency of the transition.

**Footnote:** For hydrogen, $Z = 1$; if $c$ is the velocity of light, $3 \times 10^8 \, \text{m s}^{-1}$, $R$ is the Rydberg constant, $1.097 \times 10^7 \, \text{m}^{-1}$, $h$ is Plank's constant, $6.626 \times 10^{-34} \, \text{J s}$, and $N$ is Avogadro's number, $6.022 \times 10^{23} \, \text{mol}^{-1}$, the expression becomes: $E_i - E_j = h\nu = hZ^2cR(1/n_j{}^2 - 1/n_i{}^2)$. Derived from this expression for hydrogen are a series of

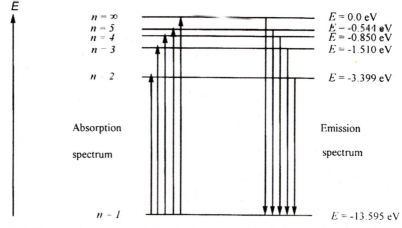

**Figure 2.4** *The relative energies of the circular orbits of the Bohr model of the hydrogen atom. The electronic energies that give rise to the line spectrum of the hydrogen atom*

named line spectra: for $n_i = 1$, Lyman; for $n_i = 2$, Balmer; for $n_i = 3$, Paschen, *etc*; each corresponding to increasing values of $n_j$ (Figure 2.5), and occurring in the far ultraviolet, the visible and the near infrared regions of the electromagnetic spectrum, respectively. Table 2.5 lists the energies in eV, and the distances from the nucleus in Å and pm.

However, although the Bohr theory, involving a single quantum number $n$, was adequate to explain the line spectrum of the hydrogen atom with a single valence electron (Figures 2.4 and 2.5, respectively), it was inadequate to explain, in detail, the line spectrum of elements with more than one electron. To do this, it was found necessary to introduce the idea of **three** further quantum numbers, in addition to the principal quantum number, $n$. These arise from the wave nature of the electron.

**Footnote: The Wave Nature of the Electron.** So far the electron has been considered as a particle, with clearly quantised energy levels, that can be precisely measured, as in the emission lines of the spectrum of hydrogen. Because the electron is so small and light, the accuracy with which it can be measured is very uncertain. This is associated with the Heisenberg Uncertainty Principle, which states that 'it is impossible to determine both the position and momentum of an electron simultaneously', *i.e.* $\Delta x \cdot \Delta p = h/2\pi$, where $\Delta x$ is the uncertainty in measuring the position of the electron and $\Delta p$ is the uncertainty in measuring the momentum ($p$ = mass × velocity) of the electron. The two uncertainties bear an inverse relationship to each other. Consequently, if the position of the

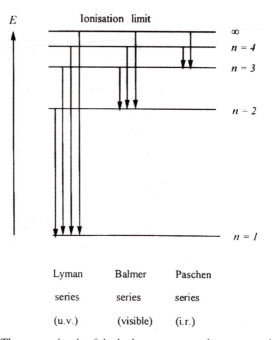

Lyman          Balmer         Paschen

series         series         series

(u.v.)         (visible)      (i.r.)

**Figure 2.5**  *The energy levels of the hydrogen atom and some named series*

**Table 2.5**  *The energies of the hydrogen K, L, M, N and O shells*

| $n$ | $2n^2$ | *Shells* | *Energy* eV | *Distance* Å | *Distance* pm ( $= 0.01$ Å) |
|----|----|----|----|----|----|
| 1 | 2 | K | − 13.595 | 0.529 | 52.9 |
| 2 | 8 | L | − 3.399 | 2.116 | 211.6 |
| 3 | 18 | M | − 1.511 | 4.761 | 476.1 |
| 4 | 32 | N | − 0.850 | 8.464 | 846.4 |
| 5 | 50 | O | − 0.544 | 13.225 | 1322.5 |
| ∞ | — | — | 0 | ∞ | ∞ |

electron is known accurately, the velocity is uncertain, and *vice versa*. Since the electron is so small and light, the very process of measuring its position or velocity is affected by the radiation that is measuring it. This results in the electron being considered both as a particle and as a packet of electromagnetic radiation. Consequently, the properties of the electron are alternatively considered as a wave and there is an alternative wave equation, called the **Schrödinger Wave Equation.** While the detailed solution of the equation is beyond the scope of this book, a number of important consequences arise from the solution. Firstly, the ability to determine the exact position of an electron has to be replaced by the **probability** (*i.e.* a 95% probability) of finding the electron at a particular position. Secondly, the idea of a single quantum number to describe the energy of an electron has to be expanded to **four** quantum numbers.

These four quantum numbers are:

(a) $n$, the principal quantum number;
(b) $l$, the azimuthal quantum number;
(c) $m_l$, the magnetic quantum number;
(d) $m_s$, the spin quantum number.

The allowed values of these quantum numbers are then:

$$n = 1, 2, 3 \ etc. \ (\text{1st, 2nd, 3rd} \ldots \text{rows});$$
$$l = +n - 1 \ldots 0;$$
$$m_l = \pm l \ldots 0 \ (\text{number} = 2l + 1; 1 \times s; 3 \times p; 5 \times d; 7 \times f);$$
$$m_s = \pm \tfrac{1}{2}.$$

The principal quantum number, $n$, is still the most important quantum number in determining the energy of an electron. The azimuthal quantum number, $l$, describes the orbital angular momentum properties, *i.e.* the average distance of the electron from the nucleus (Figure 2.6). In particular, the $l$ quantum number determines the orbital path or **shape** of an orbital. When $n = 1, l = 0$, the electron is said to occupy *one* spherically symmetrical $s$-orbital (Figure 2.7a). For $n = 2$, $l$ values of 0 and 1 are possible. The $l = 0$ value again describes a spherically symmetrical $s$-orbital, but the $l = 1$ value generates three corresponding $m_l$ values of $+ 1$, 0 and $- 1$, corresponding to *three* distinct $p$-orbitals. These three orbitals are dumb-bell in shape (Figure 2.7b), and differ only in terms of their orientations along the three $x$, $y$ and $z$ cartesian directions and are consequently labelled, $p_x$, $p_y$ and $p_z$, respectively. Likewise for $n = 3$, $l$-values of 0, 1 and 2 are possible. The $l = 0$ value generates an $s$-orbital, the $l = 1$ value generates three $p$-orbitals, while the $l = 2$ value generates *five* $d$-orbitals, which display the even more complicated dumb-bell shapes of Figure 2.7c, again with differing directional properties related to the three cartesian directions. These are labelled $d_{z^2}$, $d_{x^2-y^2}$, $d_{xy}$, $d_{xz}$ and $d_{yz}$. For $n = 3$ and $l = 3$ an additional seven $f$-orbitals arise, with even more complicated shapes, the details of which are outside the scope of this text. These results from the orbitals with differing $l$-values are summarised in Table 2.6, with the number of orbitals generated given by $2l + 1$.

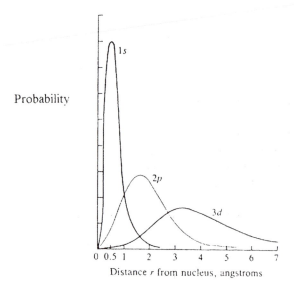

**Figure 2.6** *The probability of finding an electron at a given distance, r, from the nucleus*

**Table 2.6** *l-Values, orbital descriptions and capacity*

| l-Value | Orbital | (2l + 1) value | Total capacity |
|---------|---------|----------------|----------------|
| 0 | s-orbital | 1 | 2 electrons |
| 1 | p-orbital | 3 | 6 electrons |
| 2 | d-orbital | 5 | 10 electrons |
| 3 | f-orbital | 7 | 14 electrons |

The full set is given in Figure 2.7a–c. As each orbital generated can only hold two electrons with spin of $+\frac{1}{2}$ and $-\frac{1}{2}$ respectively, the $2l + 1$ relationship determines the total capacity for the orbitals involved, namely *two* electrons for the s-orbital, $s^2$, *six* electrons for the p-orbitals, $p^6$, *ten* electrons for the d-orbitals, $d^{10}$ and *fourteen* electrons for the f-orbitals, $f^{14}$. The relationship between the original **K, L, M** and **N** shells of the Bohr theory and the new orbital description is shown in Table 2.7.

Table 2.7 also shows how the generation of subshells, s- p-, d- and f-orbitals, results in the build-up to the valence shell configuration of an atom as each n-value generates $n - 1$ l-values, each l-value $2l + 1 m_l$ values, and each $m_l$ value two $m_s$ values of $\pm \frac{1}{2}$. This build-up process then generates the increasing capacity of the **K, L, M,** and **N** shells for

(a) The *s*-orbital.

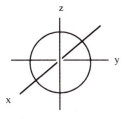

(b) The three *p*-orbitals.

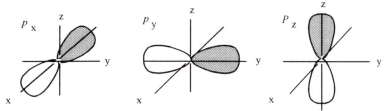

(c) The five *d*-orbitals.

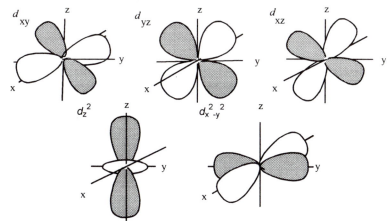

**Figure 2.7** *The shapes of* (a) *the s-orbital,* (b) *the three p-orbitals and* (c) *the five d-orbitals*

electrons: 2, 8, 18 and 32 for the *n* values of 1, 2, 3 and 4 respectively. For each principal number, *n*, the energy of the shell increases:

$$1 < 2 < 3 < 4 \text{ or } \mathbf{K} < \mathbf{L} < \mathbf{M} < \mathbf{N}$$

and within each principal quantum number *n*, the energies of the orbitals increase:

**Table 2.7** *Relationship between the shell and orbital notation*

| Shell | $n$ | $l$-Values | Orbital types | Capacity | Total |
|-------|-----|------------|---------------|----------|-------|
| **K** | 1 | 0 | 1s | 2 | 2 |
| **L** | 2 | 0, 1 | 2s, 2p | 2, 6 | 8 |
| **M** | 3 | 0, 1, 2 | 3s, 3p, 3d | 2, 6, 10 | 18 |
| **N** | 4 | 0, 1, 2, 3 | 4s, 4p, 4d, 4f | 2, 6, 10, 14 | 32 |

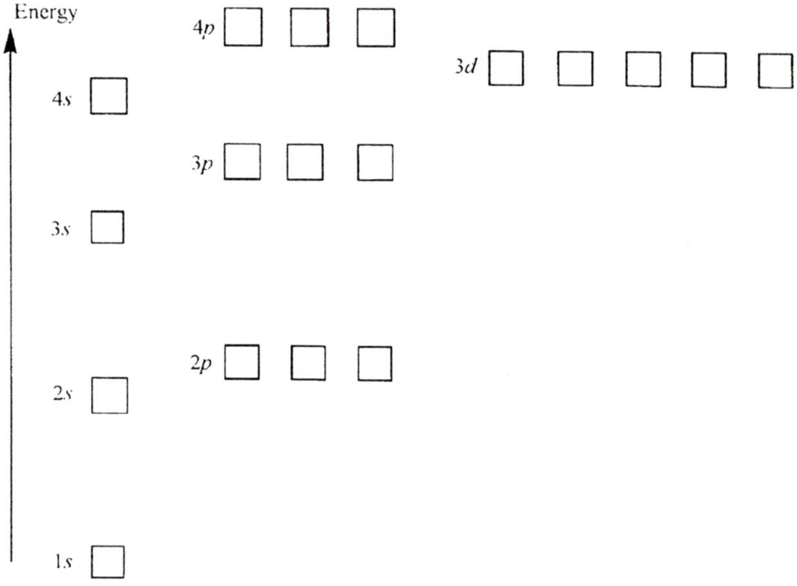

**Figure 2.8** *Energy level diagram for atomic orbitals for an atom with more than one electron*

$$s < p < d < f$$

Unfortunately, this over-simplified filling sequence does have some exceptions. The first of these is that the 3d level is slightly higher in energy than the 4s level, but lower than the 4p. Consequently, the 4s level is lower in energy than the 3d levels and fills before it. A comparable complication occurs with the 5s, 4d and 6s, 5d pairs of levels. The final energy level diagram has the form shown in Figure 2.8, where each box has a capacity for 2 electrons with $m_s$ values of $\pm \frac{1}{2}$. The degeneracy of each level is indicated by the number of boxes, namely, $2l + 1$, one box for the s-orbitals, three boxes for the p-orbitals, five boxes for the d-orbitals, and seven boxes for the f-orbitals, resulting in a total capacity

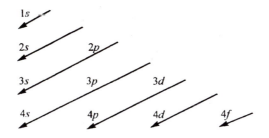

**Figure 2.9** *The order of occupancy of atomic orbitals, in the direction of the arrows from top right to bottom left*

of 2, 6, 10 and 14 electrons, respectively. An alternative scheme to show the sequence of filling of the $s$-, $p$-, $d$- and $f$-orbitals is shown in Figure 2.9.

The best evidence for the one electron orbital shell structure of the atom arises from the number of peaks (Figure 2.10) in the photo-electronic spectra of the atoms. In the H and He atoms with only single $1s$ orbital occupied, only a single peak occurs in the spectra, but that for the He $1s$-orbital occurs at a higher energy owing to the increasing nuclear charge of the He atom. In the Li atom, $1s^2 2s^1$, two peaks are observed, the higher energy one occurring owing to the $1s^2$ configuration, shifted to even higher energy owing to the increased nuclear charge of the Li atom, and a second lower energy peak due to the $2s^1$ configuration. In this two-peaked spectrum, the lower energy peak is half as intense as the higher energy one, as it only involves a single $2s^1$ electron, relative to the two electrons of the $1s^2$ configuration. Two peaks of equal intensity are observed in the spectrum of Be, $1s^2 2s^2$, while for B, $1s^2 2s^2 2p^1$, three peaks are observed with the third peak at lowest energy and with half the intensity associated with the $2p^1$ configuration, namely, 1:2:2. With Ne, $1s^2 2s^2 2p^6$, three peaks are observed with increasing intensities in the ratio 6:2:2, while with Na, $1s^2 2s^2 2p^6 3s^1$, four peaks are observed owing to the addition of the lowest energy peak associated with the $3s^1$ configuration, with intensities in the ratio 1:6:2:2.

Thus while the Bohr theory provides evidence for the principal quantum number, $n$, shell structure of the atom, the photoelectronic spectra provide the evidence for the azimuthal, $l$, and the magnetic, $m_l$, quantum numbers. The evidence for the two spin quantum numbers $\pm m_s$ was obtained by subjecting a beam of silver atoms (Ag) to a non-homogeneous magnetic field, which divided the silver atoms into two types, according to the spin angular momentum, $\pm \frac{1}{2}$, of the outer $5s^1$ configuration of the Ag atom.

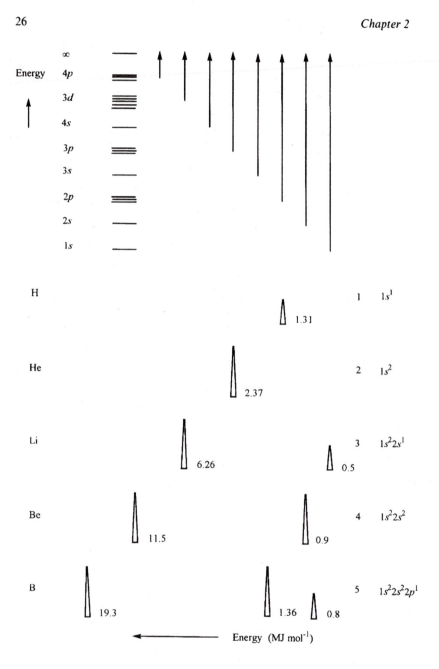

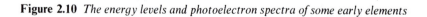

**Figure 2.10** *The energy levels and photoelectron spectra of some early elements*

**Table 2.8** *Systematic build-up process for the first ten elements of the Periodic Table*

| Element | | $1s$ | $2s$ | $2p$ | |
|---------|----|------|------|------|---|
| Hydrogen | H | $1s^1$ | | | |
| Helium | He | $1s^2$ | | | inert gas core |
| Lithium | Li | $1s^2$ | $2s^1$ | | |
| Beryllium | Be | $1s^2$ | $2s^2$ | | |
| Boron | B | $1s^2$ | $2s^2$ | $2p^1$ | |
| Carbon | C | $1s^2$ | $2s^2$ | $2p^2$ | |
| Nitrogen | N | $1s^2$ | $2s^2$ | $2p^3$ | |
| Oxygen | O | $1s^2$ | $2s^2$ | $2p^4$ | |
| Fluorine | F | $1s^2$ | $2s^2$ | $2p^5$ | |
| Neon | Ne | $1s^2$ | $2s^2$ | $2p^6$ | inert gas core |

## THE BUILD-UP PROCESS FOR THE PERIODIC TABLE

For *many electron atoms* the filling up or build-up process follows the rules:

- the lowest energy level is filled first;
- the capacity of each box is two electrons with $m_s = \pm\frac{1}{2}$ spin values;
- no two electrons may have the same values of all four quantum numbers, $n$, $l$, $m_l$ and $m_s$ (the Pauli Exclusion Principle);
- in degenerate levels, each level is half-filled before electron pairing occurs (Hund's Maximum Multiplicity Rule).

This process is illustrated in Table 2.8. In this build-up process, an $s^2$ configuration is referred to as a closed shell configuration, as the $s$-orbital is a non-degenerate level, and the total capacity of two electrons must involve opposite spin quantum numbers, $m_s$ of $\pm\frac{1}{2}$. Likewise, the $p^6$, the $d^{10}$ and the $f^{14}$ configurations are referred to as closed shell configurations, as the respective degenerate orbital levels are full. In the case of degenerate levels, such as the $2p$ levels, three orbitals are present, $2l + 1$, $0$, $-1$, and are indicated by drawing the boxes in contact *i.e.* □ □ □ for the triply degenerate $p$-orbital level, all of which have the same energy. In the case of degenerate orbitals Hund's Maximum Multiplicity Rule states that all the degenerate levels are first filled with a single electron before spin pairing occurs. Thus, the electron configuration of nitrogen $1s^2 2s^2 2p^3$ is [↑↓][↑↓][↑][↑][↑] and not [↑↓][↑↓][↑↓][↑][ ]. This latter configuration represents a higher energy level excited state. Using the energy level diagram of Figure 2.8 and the build-up process of Figure 2.9, this leads to the electron con-

**Table 2.9** *The electron configuratios of the first 18 elements*

| Atomic number | Symbol | Electron configuration | |
|---|---|---|---|
| 1 | H | $1s^1$ | |
| 2 | He | $1s^2$ | inert gas core |
| 3 | Li | $[He]2s^1$ | |
| 4 | Be | $[He]2s^2$ | |
| 5 | B | $[He]2s^22p^1$ | |
| 6 | C | $[He]2s^22p^2$ | |
| 7 | N | $[He]2s^22p^3$ | |
| 8 | O | $[He]2s^22p^4$ | |
| 9 | F | $[He]2s^22p^5$ | |
| 10 | Ne | $[He]2s^22p^6$ | inert gas core |
| 11 | Na | $[Ne]3s^1$ | |
| 12 | Mg | $[Ne]3s^2$ | |
| 13 | Al | $[Ne]3s^23p^1$ | |
| 14 | Si | $[Ne]3s^23p^2$ | |
| 15 | P | $[Ne]3s^23p^3$ | |
| 16 | S | $[Ne]3s^23p^4$ | |
| 17 | Cl | $[Ne]3s^23p^5$ | |
| 18 | Ar | $[Ne]3s^23p^6$ | inert gas core |

figuration of the elements, Table 2.9, that ultimately leads to the full Periodic Table of the elements (Figure 2.11).

In the electron configurations of Table 2.9 there is a systematic build-up process to the filled electron configurations of the inert gas core structures, namely:

$$\begin{array}{llllll} \text{Ne} & 1s^2 & 2s^2 & 2p^6 & & & [\text{Ne}] \\ \text{Ar} & 1s^2 & 2s^2 & 2p^6 & 3s^2 & 3p^6 & [\text{Ar}] \end{array}$$

which may be abbreviated as [He], [Ne] and [Ar], respectively. This results in an abbreviated electron configuration for carbon, C, $1s^22s^22p^2$ as $[He]2s^22p^2$, for sulfur, S, $1s^22s^22p^63s^23p^4$ as $[Ne]3s^23p^4$ and for iron, Fe, $1s^22s^22p^63s^23p^64s^23d^6$ as $[Ar]4s^23d^6$. This then leads naturally to a valence shell configuration for the carbon atom of $2s^22p^2$, for sulfur of $3s^23p^4$ and for iron of $4s^23d^6$, in which the closed inert gas cores of [He], [Ne] and [Ar], respectively, are omitted, on the understanding that these inert gas cores are never broken into in the chemistry of these elements and their compounds. In practice, it is, for example, the valence shell configuration of $4s^23d^6$ for iron that determines the position of iron in the Periodic Table (Figure 2.11) and determines the chemistry of its compounds.

LONG FORM of the PERIODIC TABLE of the ELEMENTS

| 1 | 2 | 3 | 4 | 5 | 6 | 7 | 8 | 9 | 10 | 11 | 12 | 13 | 14 | 15 | 16 | 17 | 18 |
|---|---|---|---|---|---|---|---|---|---|---|---|---|---|---|---|---|---|
| IA | IIA | IIIB | IVB | VB | VIB | VIIB | | VIIIB | | IB | IIB | IIIA | IVA | VA | VIA | VIIA | VIIIA |
| H | | | | | | | | | | | | | | | | | He |
| $1s^1$ | | | | | | | | | | | | | | | | | $1s^2$ |
| Li | Be | | | | | | | | | | | B | C | N | O | F | Ne |
| $2s^1$ | $2s^2$ | | | | | | | | | | | $2p^1$ | $2p^2$ | $2p^3$ | $2p^4$ | $2p^5$ | $2p^6$ |
| Na | Mg | | | | | | | | | | | Al | Si | P | S | Cl | Ar |
| $3s^1$ | $3s^2$ | | | | | | | | | | | $3p^1$ | $3p^2$ | $3p^3$ | $3p^4$ | $3p^5$ | $3p^6$ |
| K | Ca | Sc | Ti | V | Cr | Mn | Fe | Co | Ni | Cu | Zn | | | | | | |
| $4s^1$ | $4s^2$ | $3d^1$ | $3d^2$ | $3d^3$ | $3d^4$ | $3d^5$ | $3d^6$ | $3d^7$ | $3d^8$ | $3d^9$ | $3d^{10}$ | | | | | | |

**Figure 2.11** *A shortened version of the Long Form of the Periodic Table – 30 elements*

The sequence of filling of the electron sub-shells of Figure 2.8 to their individual capacities, 2 electrons for the *s*-levels, 6 electrons for the *p*-levels, 10 electrons for the *d*-levels, and 14 electrons for the *f*-levels, determines the lay out of the blocks of the Long Form of the Periodic Table. This then determines the widths of the *s*, *p*, *d* and *f* blocks for the Periodic Table, and the heights of the blocks are dependent on the principal quantum number *n*, 1–7. It is this Long Form of the Periodic Table that then summarises the periodic recurrence of the chemical properties of the elements and defines the vertical Groups I–VIII of the two short periods. More recently, the vertical Groups of the Periodic Table have been expanded to include the 10 transition metal elements, giving vertical Groups 1–18. This notation is included in Figure 2.11, but will not be used in this text.

It is in this way that the position of an element in the Periodic Table is determined by its electron configuration and can be used to predict many of the simple physical and chemical properties of the elements in their compounds.

*Chapter 3*

# The Physical Properties of the Elements and the Periodic Table

## AIMS AND OBJECTIVES

This chapter describes the connections between the one-electron configurations of the elements, the structure of the Long Form of the Periodic Table and the physical properties of the elements, namely their size, ionisation energies and electron affinities or attachment enthalpies.

## THE PERIODIC TABLE

Chapter 2 established the main features of the abbreviated Long Form of the Periodic Table, *i.e.* four horizontal rows of the elements (Figure 3.1a), determined by the principal quantum number $n$, 1–4, and eight vertical columns or groups, I–VIII, if the transition metals, lanthanides and actinides are excluded. If the 10 transition metal elements are included (Figure 3.1b), then the eight vertical groups increase to 18, a notation that will not be used in this text. The vertical groups involve a characteristic electron configuration involving an inert gas core plus an outer valence shell of electrons, *i.e.* [inert gas core] plus a valence shell configuration, $s^m p^n$. The nature of the electrons in the valence shells determines the three blocks of the Periodic Table (Figure 3.1c), namely the s-block, Groups I and II elements, the p-block, Groups III–VIII, and the d-block elements involving the series of 10 transition metal elements. Each of these blocks involves the systematic filling of the $s^1$–$s^2$, $p^1$–$p^6$ and $d^1$–$d^{10}$ electron shells, respectively, from left to right; hence the naming of the three blocks. However, within each block the principal quantum number, $n$, increases down each block, along with a corresponding increase in $Z$ (Figure 3.2).

31

Chapter 3

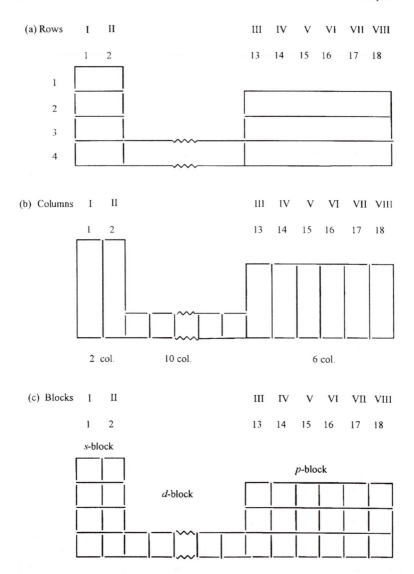

**Figure 3.1** *The Long Form of the Periodic Table*

Each block has a distinctly different chemistry, and within each block there is a more subtle variation of the chemistry depending on the valence shell electron configuration. Figure 3.3 shows an abbreviated form of the Periodic Table, Groups I–VIII, with the valence shell configurations, shown as Lewis dot structures, to emphasise the vertical

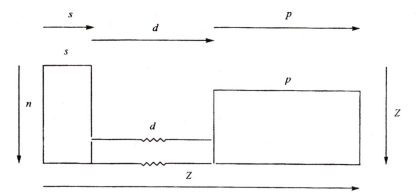

**Figure 3.2** *The variation of n, Z, s, p, and d in the Periodic Table*

| I | II | III | IV | V | VI | VII | VIII |
|---|----|-----|----|---|----|-----|------|
| 1 | 2 | 13 | 14 | 15 | 16 | 17 | 18 |

H•

$1s^1$

He :

$1s^2$

Li•   Be :   B :   • C :   • N :   • O :   • F :   : Ne :

[He] $2s^1$   $2s^2$   $2s^2 2p^1$   $2s^2 2p^2$   $2s^2 2p^3$   $2s^2 2p^4$   $2s^2 2p^5$   $2s^2 2p$

Na •   Mg:   Al:   • Si :   • P :   • S:   • Cl :   : Ar:

[Ne] $3s^1$   $3s^2$   $3s^2 3p^1$   $3s^2 3p^2$   $3s^2 3p^3$   $3s^2 3p^4$   $3s^2 3p^5$   $3s^2 3p$

K•   Ca :

[Ar] $4s^1$   $4s^2$

**Figure 3.3** *Abbreviated Periodic Table, valence shell configuration, dot form*

group relationship. Within each valence shell configuration, the chemistry will depend upon the value of the principal quantum number, $n$, and the way it influences the size of the atoms. Of particular importance are the closed valence shell electron configurations, $1s^2$, $2s^2 2p^6$ and $3s^2 3p^6$

of the inert gases He, Ne and Ar, respectively. These empty or filled shell configurations have an inherent stability in their own right, but in addition the half-filled shells, such as $p^3$, have some inherent stability.

The main variations of properties of the elements that are summarised in the Periodic Table can be divided into physical and chemical properties. These will be briefly described with the elements restricted to the first four rows of the Periodic Table in order to conserve space. The three most important physical properties of the elements are their size, ionisation potential and electron affinity or attachment enthalpy; each of these will be discussed briefly.

## VARIATION IN THE ATOMIC RADII

The variation of the atomic radii of the elements is shown in Figure 3.4, with the values given in picometres (1 pm $\equiv$ 0.01 Å). The size of an atom increases significantly down a group as the atomic number, $Z$, increases. The size of an atom decreases along a row, as, although the atomic number increases slightly, the increase in nuclear charge outways the latter. Consequently, the largest atoms are to be found at the bottom left of the Periodic Table and the smallest at the top right. Table 3.1a shows how the atomic radii of the alkali metals vary down Group I, Table 3.1b shows how the halogens vary down Group VII and Table 3.1c shows how the atomic radii varies across the second short period. Owing to the loss of electrons, cations are smaller than the parent atoms, while the anions are larger, owing to the gain of electrons. The data of Figure 3.4 and Table 3.1 are for illustrative purposes only and need not be memorised.

## VARIATION IN IONISATION POTENTIAL

The amount of energy required to remove the most loosely bound electron from a gaseous neutral atom is called the first ionisation potential

$$X_{(g)} + \text{energy} \rightarrow X^+_{(g)} + e^-$$

Table 3.2 shows the variation of the first ionisation energy for the first two short periods of the Periodic Table. In general, the energies increase across the Periodic Table, owing to decreasing size of the atom and the increasing nuclear charge. The energies decrease down the group owing to the increasing size of the atom and to the increasing 'Screening Effect' of the inner electron shells, which dilute the effect of the increasing

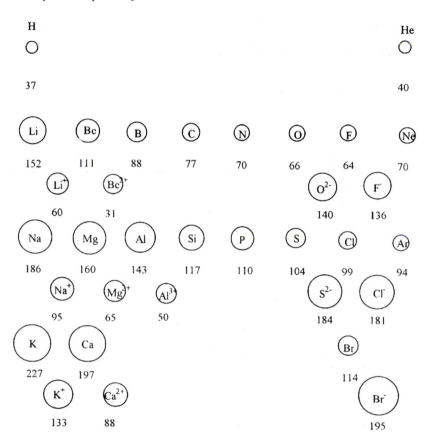

**Figure 3.4** *Variation in the atomic and ionic radii in the Periodic Table. The radii are given in picometres, with the circles not drawn to scale*

**Table 3.1** *The variation of atomic radii of the elements, (a) the alkali metals, (b) the halogens and (c) the second short period*

| (a) | The Alkali Metals | | | | | | |
|-----|-----|-----|-----|-----|-----|-----|-----|
| | Li | Na | K | Rb | | | |
| | 152 | 186 | 227 | 248 | | | |
| (b) | The Halogens | | | | | | |
| | F | Cl | Br | I | | | |
| | 64 | 99 | 114 | 133 | | | |
| (c) | Second Short Period | | | | | | |
| | Na | Mg | Al | Si | P | S | Cl |
| | 186 | 160 | 143 | 117 | 110 | 104 | 99 |

**Table 3.2** *The variation of the first ionisation potential of the elements of the first two short periods of the Periodic Table/kJ mol$^{-1}$*

| H | He | | | | | | |
|------|------|-----|------|------|------|------|------|
| 1310 | 2370 | | | | | | |
| Li | Be | B | C | N | O | F | Ne |
| 520 | 900 | 800 | 1090 | 1400 | 1310 | 1680 | 2080 |
| Na | Mg | Al | Si | P | S | Cl | Ar |
| 490 | 730 | 580 | 780 | 1060 | 1000 | 1250 | 1520 |
| K | Ca* | | | | | | |
| 420 | 590 | | *Transition metals Sc, 630–Zn, 910 | | | | |

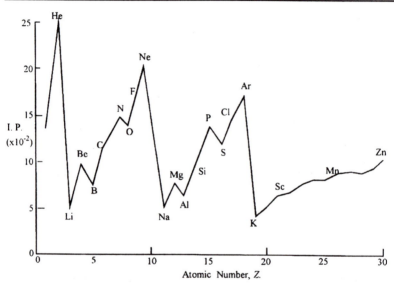

**Figure 3.5** *First ionisation potential* (kJ mol$^{-1}$) *versus atomic number, Z*

nuclear charge. The increase across Table 3.2 reflects the changes in the electron configuration $s^1 p^0 – s^2 p^6$, with the maximum reflecting the increasing stability of certain configurations such as $s^2$, $s^2 p^3$, $s^2 s^6$, *i.e.* the inherent stability of the empty, the half-filled and the completely filled subshells. This is best illustrated graphically in Figure 3.5, showing the plot of the first ionisation potential against the atomic number, $Z$. The inert gases, with closed inert gas cores, occupy the peaks and the alkali metals, with $s^1$ valence electron shells, occupy the minima of the graph. The stability of the half-filled $p^3$ configuration is then reflected in the ionisation potentials of N > O and P > S. The ionisation energies of the 10 first-row transition metals show a less significant increase with $Z$ owing to the lower shielding efficiency of a $d^n$ configuration. Successive

**Table 3.3** *Some successive ionisation potentials/kJ mol$^{-1}$*

|    | 1st  | 2nd  | 3rd   | 4th   | 5th   |
|----|------|------|-------|-------|-------|
| H  | 1312 |      |       |       |       |
| He | 2372 | 5256 |       |       |       |
| Li | 520  | 7297 | 11810 |       |       |
| Be | 899  | 1757 | 14845 | 21000 |       |
| B  | 800  | 2426 | 3659  | 25020 | 32820 |

ionisation potentials, $M^n \rightarrow M^{n+1}$, increase rapidly (Table 3.3) owing to the added attraction of the cation formed on the outermost electron.

Successive ionisation potentials that break into a lower closed inert gas core configuration show an exceptional increase and explain why these lower inert gas cores are never broken into in the chemistry of the elements.

## VARIATION IN ELECTRON AFFINITIES OR ATTACHMENT ENTHALPIES

The amount of energy required to add an electron to the lowest available empty orbital of an atom in the gaseous state is called the electron affinity or attachment enthalpy:

$$X_{(g)} + e^- \rightarrow X^-_{(g)} + \text{energy}$$

**Table 3.4** *Some electron affinities or attachment enthalpies of the elements/kJ mol$^{-1}$*

| H    | He   |      |       |      |       |       |      |
|------|------|------|-------|------|-------|-------|------|
| − 72 | + 20 |      |       |      |       |       |      |
| Li   | Be   | B    | C     | N    | O     | F     | Ne   |
| − 60 | —    | − 23 | − 123 | − 7  | − 141 | − 322 | + 30 |

Elements on the left of the Periodic Table have little tendency to add electrons, *i.e.* they have positive electron affinities. Elements to the right do accept electrons to form negative anions and hence complete an inert gas core, and hence the halide group all have high negative electron affinities, due to the formation of $X^-$ anions, with inert gas core, $s^2p^6$, electron configurations. Similarly O and S have negative electron affinities, due to the formation of inert gas core, $O^{2-}$ and $S^{2-}$ species. As the electron affinity is not an easy property to measure, only a limited amount of data is available (Table 3.4), and because of this little use is made of this physical property, but a discussion of the related term electronegativity will be considered in a subsequent section. As the values here show little systematic variation, no attempt will be made to comment on them.

## SUMMARY

It is not necessary to learn the numerical values of the various physical properties, but the trends that they follow from left to right and from the top to the bottom of the Periodic Table are important. Thus, it is useful information that with the exception of H and He, fluorine, F (64 pm), is the smallest element and potassium, K (227 pm) is the largest element in the first 18 elements of the Periodic Table. These three physical properties are clearly determined by the electron configurations of the elements and their positions in the Periodic Table. It is just this combination of these three physical properties that is responsible for the chemical properties of the elements.

*Chapter 4*

# Chemical Properties of the Elements and the Periodic Table

## AIMS AND OBJECTIVES

This chapter shows the connection between the one-electron orbital configuration of the elements and their positions in the Periodic Table and their characteristic or group oxidation numbers, their variable valances and their abilities to form ionic and covalent bonds in simple molecules.

## INTRODUCTION

Chapter 2 introduced the electronic structure of the atom and showed how the $s$-, $p$-, $d$- and $f$-orbitals generate the corresponding $s$-, $p$-, $d$- and $f$-blocks of the Long Form of the Periodic Table. Chapter 3 developed the physical properties of the elements and showed how the atomic size and first ionisation energies of the elements vary across and down the Periodic Table. In particular, the ionisation energies highlight the importance of the empty, half-filled and completely filled $s^m$ and $p^n$ orbital configurations. In order to understand the chemical consequences of these electronic properties, it is necessary to understand the origin of:

(a) Characteristic or Group Oxidation Numbers, GON;
(b) Oxidation Numbers, ON;
(c) Variable Valence, VV;
(d) Ionic and Covalent Bonding.

in the first 30 elements of the Periodic Table. The valence shell configur-

**Table 4.1**  *Abbreviated Short Form of the Periodic Table*

|        | H $s^1$ |        |          |          |          |          |          |          | He $s^2$ |
|--------|---------|--------|----------|----------|----------|----------|----------|----------|-----------|
| [He]   | Li $s^1$ | Be $s^2$ | B $s^2p^1$ | C $s^2p^2$ | N $s^2p^3$ | O $s^2p^4$ | F $s^2p^5$ | Ne $s^2p^6$ | |
| [Ne]   | Na $s^1$ | Mg $s^2$ | Al $s^2p^1$ | Si $s^2p^2$ | P $s^2p^3$ | S $s^2p^4$ | Cl $s^2p^5$ | Ar $s^2p^6$ | |
| [Ar]   | K $s^1$ | Ca $s^2$ | | | | | | | |

First Row Transition Metals:

| [Ar] $4s^2$ + | Sc $d^1$ | Ti $d^2$ | V $d^3$ | Cr $d^4(s^1d^5)$ | Mn $d^5$ | Fe $d^6$ | Co $d^7$ | Ni $d^8$ | Cu $d^9(s^1d^{10})$ | Zn $d^{10}$ |
|---|---|---|---|---|---|---|---|---|---|---|

ation of the early elements of the Abbreviated Short Form of the Periodic Table are reproduced in Table 4.1.

The electron configuration of each element is represented by:

(a) An inert gas core, *i.e.* [He], [Ne] or [Ar];
(b) Plus a valence shell configuration, *s*-block [inert gas core] $s^{1\cdot2}$; *p*-block [inert gas core] $s^2 + p^{1\cdot6}$; *d*-block [inert gas core] $s^2 + d^{1\cdot10}$.

## CHARACTERISTIC OR GROUP OXIDATION NUMBERS

How the elements behave is then determined by their ability to gain or lose electrons and their relative electronegativities with respect to adjacent elements. They cannot be measured experimentally, but Pauling defined them in terms of the relative bond energies of the atoms. An abbreviated version of Pauling's Table of Electronegativities is given in Table 4.2.

It is not necessary to learn these electronegativity values, but one should know that they increase from left to right, and from the bottom to the top of Table 4.2, such that F (4.0) is the most electronegative element and potassium, K (0.8), is the least electronegative element. As a working rule, the most electronegative elements, En > 2.5, gain electrons (electrons are attracted to them), *i.e.* $F + e \rightarrow F^-$ (this process is referred to as reduction), and the least electronegative elements, EN < 2.5, lose electrons, *i.e.* $K - e \rightarrow K^+$ (this process is referred to as oxidation). An element that gains electrons to form a negatively charged anion is described as having a negative oxidation number, with the number of electrons gained written in Roman numerals, *i.e.* $F^-$ is $-$ I. An element that loses electrons to form a positively charged cation is described as having a positive oxidation number, with the number of electrons lost written in Roman numerals, *i.e.* $Na^+$ is I.

**Table 4.2** *Pauling's electronegativities*

| | | | | | | |
|---|---|---|---|---|---|---|
| H | | | | | | |
| 2.1 | | | | | | |
| Li | Be | B | C | N | O | F |
| 1.00 | 1.5 | 2.0 | 2.5 | 3.0 | 3.5 | 4.0 |
| Na | Mg | Al | Si | P | S | Cl |
| 0.9 | 1.2 | 1.5 | 1.8 | 2.1 | 2.5 | 3.0 |
| K | Ca | | | | | |
| 0.8 | 1.1 | | | | | |

**Table 4.3** *The electron configurations, characteristic oxidation number and characteristic cations and anions of the elements of the first two short periods*

| Li | Be | B | C | | N | O | F |
|---|---|---|---|---|---|---|---|
| $s^1$ | $s^2$ | $s^2p^1$ | $s^2p^2$ | | $s^2p^3$ | $s^2p^4$ | $s^2p^5$ |
| $Li^+$ | $Be^{2+}$ | $B^{3+}$ | $C^{4+}$ | | $N^{3-}$ | $O^{2-}$ | $F^-$ |
| I | II | III | IV | | $-$ III | $-$ II | $-$ I |
| 1 | 2 | 13 | 14 | | 15 | 16 | 17 |
| $Na^+$ | $Mg^{2+}$ | $Al^{3+}$ | $Si^{4+}$ | | $P^{3-}$ | $S^{2-}$ | $Cl^-$ |
| $K^+$ | $Ca^{2+}$ | | | | | | $Br^-$ |

The number of electrons lost or gained is controlled by the valence shell configuration to form the next nearest inert gas core, thus:

$$F(s^2p^5) + 1e \rightarrow F^-(s^2p^6)(-I) \qquad \text{Reduction}$$
$$O(s^2p^4) + 2e \rightarrow O^{2-}(s^2p^6)(-II) \qquad \text{Reduction}$$
$$K(s^1) - 1e \rightarrow K^+(s^0p^0)(I) \qquad \text{Oxidation}$$
$$Ca(s^2) - 2e \rightarrow Ca^{2+}(s^0p^0)(II) \qquad \text{Oxidation}$$

In this way the oxidation number table is built up (Table 4.3). These oxidation numbers are then referred to as the **Characteristic or Group** oxidation numbers of the elements and relate directly to the electron configuration of the elements and its tendency to lose or gain electrons to form an inert gas core electron configuration. As the Roman numeral notation I–VII more closely equates with the Group electron configurations, this Group oxidation number notation will be retained where is emphasises this connectivity. The positive oxidation numbers of the first four columns equate with the Group numbers of the Periodic Table, while the last three columns equate with minus 8 plus the Group number, thus for nitrogen, $-8 + 5 = -3$ or $-$ III. The cations and anions of Table 4.3 are then the stable ions of their respective groups, each having a filled inert gas core; the cations, by loss of electrons, all involve the [He], [Ne] or [Ar] inert gas cores, the anions, by gain of electrons, the [Ne] and [Ar] inert gas cores. These characteristic ions are then unique to the vertical Group and are reluctant to undergo any variable valence as ions, *i.e.* all the Group I alkali metals only form monovalent cations and all the Group VII halogens form monovalent anions. In both series the size of the ions increase down the group. When atoms lose an electron to form a cation, the cations are smaller than the free element and when atoms gain an electron to form anions, the anions are larger than the free element. These differences are shown in Table 4.4.

**Table 4.4** *The relative sizes of cations and anions/pm*

| | | | | | |
|---|---|---|---|---|---|
| Li | $Li^+$ | | | F | $F^-$ |
| 152 | 60 | | | 64 | 136 |
| Na | $Na^+$ | | | Cl | $Cl^-$ |
| 186 | 95 | | | 99 | 181 |
| K | $K^+$ | | | Br | $Br^-$ |
| 227 | 133 | | | 114 | 195 |
| | Fe | $Fe^{2+}$ | | $Fe^{3+}$ | |
| | 232 | 152 | | 128 | |

**Table 4.5** *Some examples of the stoichiometry of simple ionic salts*

| | | | |
|---|---|---|---|
| AB | NaCl | MgO | AlN |
| $AB_2$ | $MgCl_2$ | $Li_2O$ | $Na_2S$ |
| $AB_3$ | $AlF_3$ | $Na_3N$ | $Li_3N$ |
| $A_2B_3$ | $Mg_3N_2$ | $Al_2O_3$ | $Be_3P_2$ |

**Table 4.6** *The sizes of a series of isoelectronic ions, with the [Ne] inert gas core structure*

| $N^{3-}$ | $O^{2-}$ | $F^-$ | Ne | $Na^+$ | $Mg^{2+}$ | $Al^{3+}$ |
|---|---|---|---|---|---|---|
| 171 | 140 | 136 | 112 | 95 | 65 | 50 |

The relative charges of these inert gas core ions then determine the relative stoichiometry of the neutral ionic compounds formed, namely, AB, $AB_2$, $AB_3$, $A_2B_3$, *etc.* Some examples of such simple ionic compounds or salts are shown in Table 4.5. In these ionic salts, the inert gas cores indicate clear oxidation states that equate with the charges on the cations and anions and these must sum to zero, *i.e.* in $Al_2O_3$, $Al^{III}_2O^{-II}_3$ and $(+ 3 \times 2) + (- 2 \times 3) = 0$. The presence of alkali metal or alkaline earth cations in compounds such as $Na_2CO_3$ or $CaCO_3$ implies that the $CO_3$ group should be written as a discrete divalent anion, $CO_3^{2-}$, *i.e.* the carbonate anion.

The relative effect of increasing charge is also shown in the Fe, $Fe^{2+}$ and $Fe^{3+}$ series of Table 4.4. The relative effect of increasing atomic number, $Z$, is shown in the series of isoelectronic ions, *i.e.* with the same inert gas core of [Ne] (Table 4.6).

## OXIDATION NUMBERS

The simple origin of the Characteristic or Group oxidation numbers described above suggests that it only applies to simple anions and cations involving ionic bonding as in, for example, NaCl, $Na^+Cl^-$, sodium(I). It can also be applied to a wider range of compounds, if it is recognised that the oxidation numbers are only approximations. Thus while $BF_3$ is a colourless, gaseous molecule, it may be described as boron(III) trifluoride($-I$) to reflect the origin of the formal oxidation numbers. Likewise the carbonate oxyanion, $CO_3^{2-}$, may be formally written as $C^{IV}O^{-II}_3{}^{2-}$.

## RULES FOR THE DETERMINATION OF OXIDATION NUMBERS

To determine the oxidation number (ON) of an element the following rules are used:

1. The oxidation number of a free element is zero, *e.g.* Na, B, $F_2$, $O_2$, $P_4$, *etc.*
2. The oxidation number of any monoatomic ion is equal to the charge on the ion, *e.g.* $Na^+$(I); $F^-$($-I$); $Al^{3+}$(III); $O^2$($-II$); $N^{3-}$($-III$), *etc.*
3. The sum of the oxidation number of the atoms in a molecule or ion must equal the charge on that molecule or ion, *e.g.* $KMnO_4$: $K^IMn^{VII}O^{-II}_4$; $(+1+7)-(2\times 4)=0$; $SO_4^{2-}$; $S^{VI}O^{-II}_4$; $+6-(2\times 4)=-2$.
4. In covalent hydrides, hydrogen has an oxidation number of I, but in ionic hydrides such as NaH, hydrogen has an ON of $-I$. Fluorine generally has an oxidation number of $-I$ and oxygen an ON of $-II$ (or $-I$ in peroxides such as $H_2O_2$).

The use of formal oxidation numbers can also be applied to coordination complexes such as $[Mn^{II}(OH_2)_6]^{2+}$ and $[Fe^{II}(OH_2)_6]^{2+}$ or $[Fe^{III}(OH_2)_6]^{3+}$, where the water molecule is considered as a neutral species or written as $O^{-II}H^I_2$, an overall neutral molecule.

## MAIN GROUP VARIABLE VALENCE

The idea of a characteristic oxidation number based on the electron configuration of an element suggests that the elements only form a single characteristic oxidation number, as in Figure 4.1.

Although this is generally true for the Group I(1), II(2), and III(13) elements, the Group IV–VII (14–17) elements not only gain electrons to form the characteristic oxidation states (reduction), but can also lose electrons (oxidation) to form the next lower inert gas core, as in Figure 4.2. In each case the difference between the two oxidation states $s^2p^6$ and $s^0p^0$ is 8 electrons. The lower oxidation states obtained by **addition** of

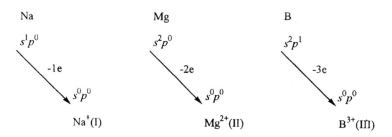

**Figure 4.1** *Group oxidation numbers (Groups I–III)*

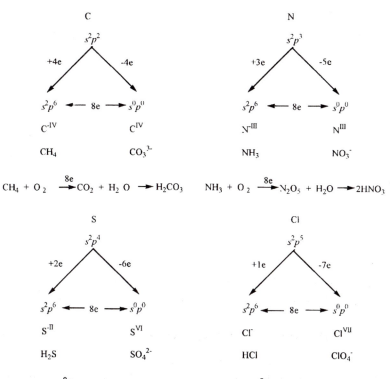

**Figure 4.2** *Group oxidation numbers (Groups IV–VIII)*

electrons (reduction) are represented by the simple hydrides of the elements and the positive oxidation states obtained by **removal** of electrons are represented by the oxides and hence oxyacids of the elements. This variable valence of the main group elements is illustrated by oxidation with molecular oxygen (Figure 4.2).

In aqueous solution the variable valence of the main group compounds is much more limited. The group I–III (1, 2, 13) elements show no variable valence (Figure 4.1). With the Group IV–VII (14–17) elements, variable valence is generally associated with the positive Group oxidation states involving a two-electron reduction process from the oxidation state and associated with the inherent stability of the $s^2$ electron configuration, the *pseudo* inert pair effect:

$$CO \quad \rightarrow CO_2 + 2e$$
$$\text{II} \qquad\quad \text{IV}$$
$$s^2p^0 \qquad s^0p^0$$

$$NO_2^- \rightarrow NO_3^- + 2e$$
$$\text{III} \qquad\quad \text{V}$$
$$s^2p^0 \qquad s^0p^0$$

$$PO_3^{3-} \rightarrow PO_4^{3-} + 2e$$
$$\text{III} \qquad\quad \text{V}$$
$$s^2p^0 \qquad s^0p^0$$

$$SO_3^{2-} \quad SO_4^{2-} + 2e$$
$$\text{IV} \qquad\quad \text{VI}$$
$$s^2p^0 \qquad s^0p^0$$

$$ClO_3^- \rightarrow ClO_4^- + 2e$$
$$\text{V} \qquad\quad \text{VII}$$
$$s^2p^0 \qquad s^0p^0$$

Figure 4.3 summarises this Main Group variable valence in a graphical form, emphasising the variation of the $s^n p^m$ configuration against the oxidation number, for the respective first and second row elements.

## TRANSITION METAL VARIABLE VALENCE

In the variable valence of the transition metals, with a $[Ar]4s^2 3d^n$ configuration, the distribution of the oxidation states (Figure 4.4) is not as systematic as in the Main Group elements (Figure 4.3). Thus, for the first-row transition metal ions the following generalisations may be made:

(a) all the oxidation states are *positive* from I to VII;
(b) in oxidation numbers from I up, the transition metal ions only contain *d*-electrons in their valence shell configuration, *i.e.* $Fe^{II}$, $[Ar]3d^6$; $V^{II}$, $[Ar]3d^3$; $Cu^{II}$, $[Ar]3d^9$;

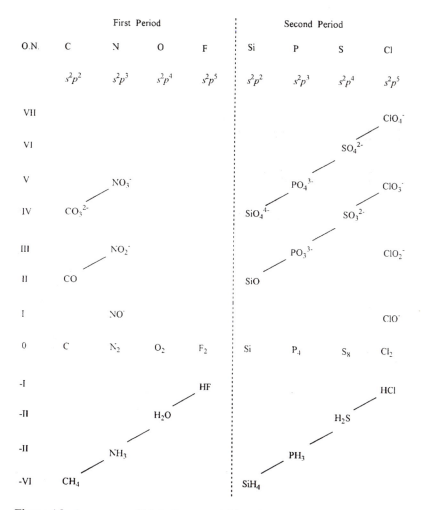

**Figure 4.3** *A summary of Main Group variable valence*

    (c) the low oxidation numbers, II–IV, are reducing agents, and the high oxidation numbers IV–VII are oxidising agents;

    (d) the oxidation numbers II–III occur as hexaaquo complexes, but the oxidation numbers of IV–VII only occur as oxyanion complexes;

    (e) in aqueous solution the redox half-reactions of the transition metal ions, involve transfers of 1–5 electrons, and frequently involve a **one** electron transfer, as in $Fe^{II} \rightleftharpoons Fe^{III}$ (in contrast to the

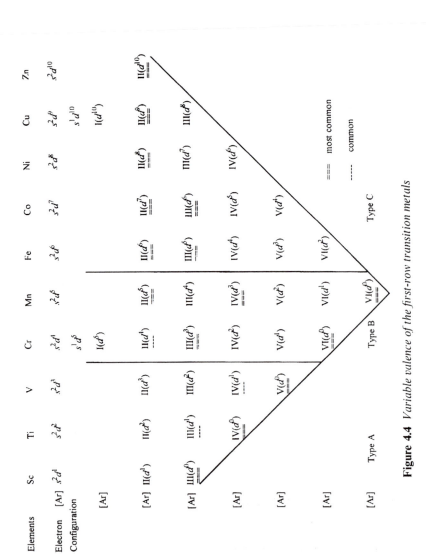

**Figure 4.4** *Variable valence of the first-row transition metals*

two-electron transfer in Main Group variable valence, *i.e.* $N^{III}O^- \xrightarrow{-2e} N^V O_3^-$;

These generalisations may be illustrated by the following half reactions:

*Oxidising agents:*

$$Mn^{VII}O_4^- + 5e \rightarrow [Mn^{II}(OH_2)_6]^{2+} \text{ (acid)}$$
$$Cr^{VI}_2O_7^{2-} + 6e \rightarrow 2[Cr^{III}(OH_2)_6]^{3+}$$

*Reducing agents:*

$$[Fe^{II}(OH_2)_6]^{2+} - 1e \rightarrow [Fe^{III}(OH_2)^6]^{3+}$$
$$[V^{II}(OH_2)_6]^{2+} - 2e \rightarrow [V^{IV}O(OH_2)_5]^{2+}$$
$$[V^{II}(OH_2)_6]^{2+} - 3e \rightarrow V^V O_3^-$$
$$[Mn^{II}(OH_2)_6]^{2+} - 2e \rightarrow Mn^{IV}O_2 \text{ (neutral)}$$
$$[Co^{II}(OH_2)_6]^{2+} - 1e \rightarrow [Co^{III}(O_2)_6]^{3+}$$
$$Cu^{I}(aq) - 1e \rightarrow [Cu^{II}(OH_2)_6]^{2+}$$

Figure 4.4 gives a summary of the variable valence of the first-row transition metals as a function of their $d^n$ configuration and oxidation number.

## CHEMICAL STOICHIOMETRY

The stoichiometry of a general reaction

$$v_A A + v_B B \rightarrow v_C C + v_D D$$

is determined by the ratio of the stoichiometry factors, $v_A : v_B$, some examples of which are given in Figure 4.5, divided in terms of the three types of reaction met in volumetric chemistry. In these three types of reactions I–III it is essential to identify correctly the nature of the reaction involved, namely (a) acid/base; (b) precipitation, or (c) redox, and then to identify the correct half-reactions, as it is this that determines the number of reactive species in each of the reactants. For this reason, the next pages summarise some useful reactions and half-reactions, with given stoichiometry factors $v_A$ and $v_B$, as these are **all** that is required to determine the stoichiometry factors in an unknown reaction:

(a) Acid/base:

$$H^+ + OH^- \rightarrow H_2O$$

I    Acid / Base reactions:                                                    $v_A : v_B$

| | | | |
|---|---|---|---|
| 1 NaOH + 1 HCl | = | 1 NaCl + 1 $H_2O$ | 1:1 |
| 1 Ba(OH)$_2$ + 2 HCl | = | 1 BaCl$_2$ + 2 $H_2O$ | 1:2 |
| 2 Al(OH)$_3$ + 3 $H_2SO_4$ | = | 1 Al$_2$(SO$_4$)$_3$ - 6 $H_2O$ | 2:3 |
| 1 Na$_4$ [Cu(OH)$_6$] + 3 $H_2SO_4$ | = | 1 CuSO$_4$ + 6 $H_2O$ + 2 Na$_2$SO$_4$ | 1:3 |

II    Precipitation reactions:

| | | | |
|---|---|---|---|
| 1 NaCl + 1 AgNO$_3$ | = | 1 AgCl + NaNO$_3$ | 1:1 |
| 1 AlBr$_3$ + 3 AgNO$_3$ | = | 3 AgBr + Al(NO$_3$)$_3$ | 1:3 |
| 1 KAlI$_4$ + 4 AgNO$_3$ | = | 4 AgI + Al(NO$_3$)$_3$ + KNO$_3$ | 1:4 |

III    Redox reactions:

| | | | |
|---|---|---|---|
| 1 [Fe(OH$_2$)$_6$]$^{3+}$ + 1 I$^-$ | = | 1 [Fe(OH$_2$)$_6$]$^{2+}$ + I$^0$ | 1:1 |
| 2 MnO$_4^-$ + 5 C$_2$O$_4^{2-}$ | = | 2 [Mn(OH$_2$)$_6$]$^{2+}$ + 10 CO$_2$ | 2:5 |
| 4 MnO$_4^-$ + 5 [Cu(C$_2$O$_4$)$_2$]$^{2-}$ | = | 4 [Mn(OH$_2$)$_6$]$^{2+}$ + 20 CO$_2$ | 4:5 |
| 12 MnO$_4^-$ + 5 [Fe(NO$_2$)$_6$]$^{3-}$ | = | 12 [Mn(OH$_2$)$_6$]$^{2+}$ + 30 NO$_3^-$ | 12:5 |
| 7 Cr$_2$O$_7^{2-}$ + 6 [Fe(SO$_3$)$_3$]$^{4-}$ | = | 14 [Cr(OH$_2$)$_6$]$^{3+}$ + 6 [Fe(OH$_2$)$_6$]$^{3+}$ | 7:6 |
| | | + 18 SO$_4^{2-}$ | |

**Figure 4.5** *Chemical stoichiometry – types of reactions*

(b) Precipitation:

$$Ag^+ + Cl^- \rightarrow AgCl \downarrow$$
$$Ag^+ + Br^- \rightarrow AgBr \downarrow$$
$$Ag^+ + I^- \rightarrow AgI \downarrow$$
$$Ba^{2+} + SO_4^{2-} \rightarrow BaSO_4 \downarrow$$
$$Ba^{2+} + SO_3^{2-} \rightarrow BaSO_3 \downarrow$$
$$Ca^{2+} + CO_3^{2-} \rightarrow CaCO_3 \downarrow$$
$$Ca^{2+} + C_2O_4^{2-} \rightarrow CaC_2O_4 \downarrow$$

(c) Redox – (I) Oxidising agents:

$$MnO_4^- + 5e \rightarrow [Mn(OH_2)_6]^{2+}$$
$$Cr_2O_7^{2-} + 6e \rightarrow 2[Cr(OH_2)_6]^{3+}$$
$$I^0 + 1e \rightarrow I^-$$

(II) Reducing Agents:

$$[Fe(OH_2)_6]^{2+} \rightarrow [Fe(OH_2)_6]^{3+} + 1e$$
$$NO_2^- \rightarrow NO_3^- + 2e$$
$$PO_3^{3-} \rightarrow PO_4^{3-} + 2e$$
$$SO_3^{2-} \rightarrow SO_4^{2-} + 2e$$
$$ClO_3^- \rightarrow ClO_4^- + 2e$$
$$I^- \rightarrow I^0 + 1e$$
$$C_2O_4^{2-} \rightarrow 2CO_2 + 2e$$

## The Calculation of Chemical Stoichiometry Factors –
## Worked Examples

In order to calculate the stoichiometry factors, $v_A$ and $v_B$, for a general reaction: $v_A.A + v_B.B \rightarrow$, it is necessary to use the Working Method of Table 4.7. Using this Working Method (Table 4.7), this identifies the number of reactive species, $e_A$ and $e_B$, for the reactants A and B, respectively. Then from the equality $v_A.e_A = v_B.e_B$, the values of $v_A$ and $v_B$ can be evaluated by inspection. For example:

1. Acid base: $v_A \cdot A + v_B \cdot B \rightarrow$
   $$Al(OH)_3 + H_2SO_4 \rightarrow$$
   Identify reactive species, *i.e.* $3OH^- (e_A = 3)$ $2H^+ (e_B = 2)$
   Identify relevant reaction, $OH^- + H^+ \rightarrow H_2O$
   $$v_A \cdot 3 = v_B \cdot 2$$
   $$v_A = 2; v_B = 3$$
   $$\therefore 2Al(OH)_3 + 3H_2SO_4 \rightarrow$$

**Table 4.7** *A Working Method to determine the stoichiometry factors* $v_A : v_B$
*for the reaction* $v_A.A + v_A.B \rightarrow v_C.C + v_D.D$.

1. Identify the two reacting molecules A and B.
2. Identify the type of reaction between A and B, *i.e.* acid/base, precipitation or redox.
3. Identify the relevant half reactions.
4. Identify the number of separate reactive species, $e_A$ and $e_B$.
5. From the equality $v_A.e_A = v_B.e_B$, solve for $v_A$ and $v_B$, by inspection.
6. If required solve for $v_C$ and $v_D$.

2. Precipitation:    $v_A \cdot A + v_B \cdot B \rightarrow$
$$PCl_5 + Ag_3PO_4 \rightarrow$$
Identify reactive species, *i.e.* $5Cl^- (e_A = 5)\ 3Ag^+ (e_B = 3)$
Identify relevant reaction, $Ag^+ + Cl^- + \rightarrow AgCl \downarrow$
$$v_A \cdot 5 = v_B \cdot 3$$
$$v_A = 3;\ v_B = 5$$
$$\therefore\ 3PCl_5 + 5Ag_3PO_4 \rightarrow$$

3. Redox:    $v_A \cdot A + v_B \cdot B \rightarrow$
$$KMnO_4 + K_4[Fe(NO_2)_6$$
Identify oxidation numbers and half reactions:
$$KMn^{VII}O_4\ K_4[Fe^{II}(N^{III}O_2)_6]$$
$Mn^{VII} + 5e \rightarrow Mn^{II}$; $Fe^{II} - 1e \rightarrow Fe^{III}$; $N^{III}O_2^- - 2e \rightarrow N^VO_3^-$
$$5e\,(e_A = 5)\ 1e + (2e \times 6) = 13e\,(e_B = 13)$$
$$v_A \cdot 5 = v_B \cdot 13$$
$$v_A = 13;\ v_B = 5$$
$$\therefore\ 13KMnO_4 + 5K_4[Fe(NO_2)_6] \rightarrow$$

## Redox Reactions

In general, these oxidation processes occur readily in aqueous solutions, where oxyanions are involved. In these oxyanion half-reactions the permanganate ion may be used as the oxidising agent:

oxidation $5 \times\ -2e$
$$5NO_2^- + 2MnO_4^- \rightarrow 5NO_3^- + 2[Mn(OH_2)_6]^{2+}$$
$2 \times\ +5e$ reduction

All of these species involve oxyanions except the $Mn^{2+}$ cation, which is stabilised as a hexaaquo cation, $[Mn^{II}(OH_2)_6]^{2+}$. In these two half-reactions the number of electrons involved in the oxidation process

( − 10e) balances the number of electrons involved in the reduction process ( + 10e).

## COVALENT BONDS

To date, the emphasis has been on the formation of ionic cations or anions, by the formation of inert gas core configurations, which then combine to form purely electrostatic bonds, *e.g.* $Na^+Cl^-$. An alternative type of bond is the covalent bond, which is characterised by the sharing of two electrons by two atoms, in a way that completes the inert gas core of both atoms. Thus, in the case of two hydrogen atoms, both with the same valence shell configuration, $1s^1$, the formation of a homonuclear diatomic molecule of $H_2$ can be represented, as follows:

$$H + H \rightarrow H — H$$

in which both H atoms have a complete $s^2$ configuration, but the electron pair is equally shared between the two H-atom centres. Such a shared pair of electrons is described as a covalent bond or a stick bond. A comparable covalent bond can be formed in $F_2$:

$$F + F \rightarrow F — F$$

but now the shared two-electron pair bond is part of the $s^2p^6$, [Ne] inert gas core of both F atoms. This type of shared two-electron pair bond, is referred to as a **single covalent bond**. As the covalent bond is between the same types of atom, the sharing is equal and the bond is referred to as a single bond and is represented as a single stick, in a homonuclear diatomic molecule. Such molecules will have no **dipole moment**.

A single covalent bond may also be formed in HF, thus:

$$H + F \rightarrow H — F$$

but in the heteronuclear single covalent bond, the shared electron pair is not shared equally, as the two atoms involved in the bond are not the same, and the bond will have a dipole moment, the sense of which will be determined by the relative **Electronegativities** (Pauling's) of the two bonded atoms (Table 4.2).

In HF, as the F atom is the most electronegative element in the Periodic Table, the F atom will involve the larger share of the electron pair and the less electronegative H atom the smaller share of the electron pair. In this heteronuclear diatomic HF molecule a dipole moment will be present:

$$H \cdots F \qquad H^{\delta+} \cdots F^{\delta-} \qquad H^+ \quad F^-$$

covalent bond     polar bond     ionic bond

The actual polar bond in HF is the intermediate between the pure covalent bond and the ionic bond, where complete transfer of an electron has occurred.

### Polyatomic Covalent Molecules

Some simple covalent polyatomic molecules are: $CH_4$, $NH_3$, $OH_2$ and HF. In the case of methane:

$$CH_4 \equiv C\text{-}s^2p^2 + 4H\text{-}s^1$$
Total electrons $4 + (4 \times 1) \equiv 8$ electrons $\equiv s^2p^6$

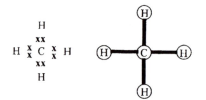

Each of the four H atoms involves a shared $s^2$ [He], inert gas configuration and the one C atom involves a shared $s^2p^6$ configuration of four electron pairs shared with four separate H atoms, which may be represented as the dots and x's in a Lewis structure or as four electron pair stick bonds.

In the case of ammonia:

$$NH_3 \equiv N\text{-}s^2p^3 + 3H\text{-}s^1$$
Total electrons $5 + (3 \times 1) \equiv 8$ electrons $\equiv s^2p^6$

Again each of the three H atoms involves a shared $s^2$ [He], inert gas configuration and the one N atom involves a shared $s^2p^6$ configuration of four electron pairs, of which three are shared with three separate H atoms plus one unshared: the latter is referred to as a lone pair of electrons.

In the same way the octet in water involves two bonding pairs of electrons and two lone pairs of electrons about the O atom, while in HF, the octet around the F atom involves one bonding pair of electrons and three lone pairs:

In this way, the valence shell configuration of the central atom, combined with the Lewis representation of the inert gas shell, gives a very useful way of visualising the distribution of the valence shell electrons in this chemical book-keeping exercise. In these Lewis structures all the electrons are equivalent and the dot or cross notation simply indicates the source of the electrons from the central atom or the terminal atoms.

A slightly different situation arises in $BH_3$:

$$BH_3 \equiv B\text{-}s^2p^1 + 3H\text{-}s^1$$
$$\text{Total electrons} \equiv 3 + 3 \equiv 6 \equiv s^2p^4$$

Here the final valence shell electron configuration is less than eight electrons and as a consequence the $BH_3$ molecule can attract a further two electrons to complete its octet, by accepting a share of the lone pair of electrons from the $NH_3$ molecule to form a stable addition compound, $H_3BNH_3$:

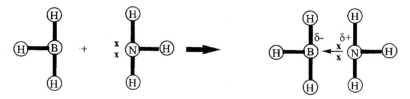

In this addition compound, the N to B bond is referred to as a **dative covalent bond**, as although the two-electron pair bond is indistinguishable for a normal shared electron pair bond, the source of the electron pair is from one of the atoms of the bond and not equally from both.

This donor function of a lone pair of electrons is quite common and occurs in coordination complexes of the transition metal ions, such as $[Ni(NH_3)_6]SO_4$, $[Mn(OH_2)_6]SO_4$ and $[Fe(OH_2)_6]SO_4$. In such complexes the $NH_3$ and $OH_2$ ligands share their lone pairs with the respective metal cations by dative covalent bond formation, a function in which they are said to act as ligands. In both of these complexes, the six-coordinated metal cations are referred to as coordination complexes, which are significantly stabilised by this process and can maintain their identity, even when dissolved in water.

It has already been shown that the $F_2$ molecule involves a single covalent bond. If the two O atoms of $O_2$ are treated in the same way they have insufficient valence shell electrons to form stable octets by sharing just one electron pair, but if two electron pairs are shared, stable octets are formed on both O atoms, with the sharing of the two electron pairs. This situation is referred to as a **double bond:**

$$\overset{xx}{\underset{xx}{\overset{x}{\underset{x}{O}}}} + \overset{xx}{\underset{xx}{\overset{}{O}\overset{x}{\underset{x}{}}}} \longrightarrow \overset{x}{\underset{x}{\overset{}{O}}}\overset{x\ x}{\underset{x\ x}{\overset{}{}}}\overset{x}{\underset{x}{\overset{}{O}}} \quad \text{or} \quad O\!=\!\!=\!O$$

In the same way the $N_2$ molecule shares three electron pairs between the two N-atoms to form a **triple bond:**

$$\overset{x}{\underset{x}{\overset{x}{N}}}x + x\overset{x}{\underset{x}{N^x_x}} \longrightarrow \overset{x}{\underset{x}{}}N\overset{x}{\underset{x}{\overset{x}{x}}}N\overset{x}{\underset{x}{}} \quad \text{or} \quad N\!\equiv\!N$$

In this way, the Lewis structures of these diatomic molecules may be used to understand the formation of single, double and triple bonds between the atoms in these molecules.

## MOLECULAR ORBITAL THEORY OF DIATOMIC MOLECULES

In ionic species like NaCl, the energy that holds the lattice together is the electrostatic attraction of the $Na^+$ cations and the $Cl^-$ anions. In covalent molecules with electron pairs or stick bonds, it is less obvious what holds the molecule together. To appreciate this it is necessary to look at the **Molecular Orbital Theory** (MO) of simple diatomic molecules. Table 4.8 shows some experimental data for some simple diatomic molecules. What is immediately noticeable about these data is that although there is a linear increase in the total number of valence shell electrons along this series of diatomic species, neither their interatomic distances nor their heats of formation increase linearly. In fact, the shortest distance and the highest heat of formation occur for the

**Table 4.8** *Some experimental properties of some simple diatomic species*

| Diatomics | $H_2^+$ | $H_2$ | $He_2^+$ | $He_2$ |
|---|---|---|---|---|
| A–A distance/pm | 106 | 74 | 108 | — |
| $\Delta H/\text{kJ mol}^{-1}$ | 255 | 430 | 251 | — |
| Number of valence electrons | 1 | 2 | 3 | 4 |

$H_2$ molecule and the $He_2$ diatomic molecule does not even exist, as helium prefers to exist as a monatomic gas, with two stable $s^2$ valence shell configurations. To understand these chemical facts it is necessary to understand the simple ideas of the **Molecular Orbital theory**, involving two $s$-orbitals on two separate H atoms, at a distance $r$ pm apart (Figure 4.6). When $r \gg 100$ pm there is no interaction between these spherically symmetrical $s$ orbitals, but as $r$ decreases to $\sim 100$ pm the outer surfaces of the $s$-orbitals (within which there is a 95% probability of finding an electron) start to interact and to occupy the same volume in space and are said to overlap (Figure 4.6). In this process the potential energy of the electrons in these $s$-orbitals will also start to interact (Figure 4.7), as the electrons in the separate $s$-orbitals have the same negative charge and repel each other. If the spins are parallel the

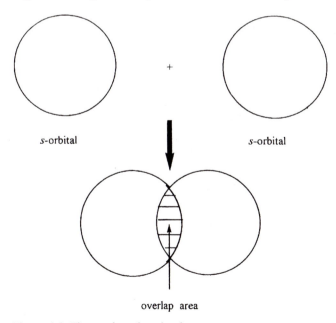

s-orbital     +     s-orbital

overlap area

**Figure 4.6** *The overlap of s-orbitals*

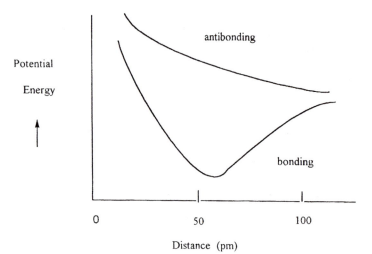

**Figure 4.7** *Potential energy versus the distance between two s-electrons*

repulsion continues and the overlap forms an *anti-bonding* molecular orbital with higher potential energy than the starting *s*-orbitals. On the other hand, if the spins of the separate electrons in the separate *s*-orbitals are opposite in sign, then the electrons are spin paired and a *bonding* molecular orbital is formed with a lowering of the potential energy. This lowering of the potential energy continues until a nuclear separation of *ca.* 75 pm (Table 4.8) occurs; thereafter, the potential energy rises, largely owing to the close approach of the two positively charge nucleus which then repel each other. With two electrons in the *bonding* molecular orbital with their spins paired, an electron pair covalent bond is formed which is stabilised relative to the isolated *s*-orbital levels by the energy $2\Delta E$ (Figure 4.8).

This molecular orbital diagram involves increasing energy vertically and the MOs have a total capacity for four electrons, with the electrons occupying the lowest energy molecular orbital first. As the individual *s* atomic orbitals, A(1) and A(2), have the same energies and a capacity of two electrons each, $m_s = \pm \frac{1}{2}$, thus providing a total capacity of four electrons. Consequently, the molecular orbitals formed must also be non-degenerate, having a capacity of two electrons each and a total capacity of four electrons. This requires that there will be two molecular orbitals formed, namely, a lower energy bonding molecular orbital, $\sigma^b$, and a higher energy anti-bonding molecular orbital, $\sigma^*$. As the energies of these molecular orbitals are relative to the starting atomic orbitals, the energy of the bonding molecular orbital is stabilised ($+\Delta E$) by approximately, the same amount of energy, as the anti-bonding molecu-

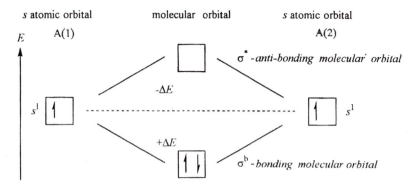

**Figure 4.8** *The molecular orbital diagram of the $H_2$ diatomic molecule*

**Table 4.9** *The electron occupancy of the molecular orbital diagram of Figure 4.6*

|  | A1 | A2 | $\sigma^b$ | $\sigma^*$ | Energy | BO |
|---|---|---|---|---|---|---|
| $H_2^+$ | 1 | 0 | 1 | 0 | $\Delta E$ | 0.5 |
| $H_2$ | 1 | 1 | 2 | 0 | $2\Delta E$ | 1.0 |
| $He_2^+$ | 2 | 1 | 2 | 1 | $\Delta E$ | 0.5 |
| $He_2$ | 2 | 2 | 2 | 2 | 0 | 0.0 |

lar is destabilised ( $-\Delta E$). It is for this reason that the lower molecular orbital, $\sigma^b$, containing two electrons, is referred to as a **bonding** molecular orbital, as there is a gain of energy to form a two-electron pair covalent bond. Equally, the higher molecular orbital, $\sigma^*$, is referred to as an **anti-bonding** molecular orbital, as it is *against* bonding, from an energy point of view. In the multiple electron species, 1–4, the lowest energy MO is occupied first and the filling process of Table 4.9 results. In this way, while electrons in the $\sigma^b$ orbital contribute to bonding, with an energy of $+\Delta E$, those in the $\sigma^*$ orbital are not only against bonding, but electrons in this anti-bonding level, with an energy of $-\Delta E$, cancel out the bonding effect of electrons in the bonding level. Thus in $He_2$, the two electrons in the bonding molecular orbital, $\sigma^b$, are cancelled out by the two electrons in the anti-bonding molecular orbital, $\sigma^*$; this results in the net no-bonding for the diatomic $He_2$ species, consistent with the non-existence of this molecule and the occurrence of helium as a mon-atomic gas. Thus, this simple use of the molecular orbital theory not only explains the source of energy for the covalent bond, but also accounts for some simple chemistry, *i.e.* the existence of the diatomic $H_2$ molecule and the non-existence of the corresponding $He_2$ species. It also explains the comparable interatomic distances of the $H_2^+$ and $He_2^+$ cationic species, both greater than that in $H_2$, and the comparable $\Delta H$

values of these two cations, both of which are larger than that of the $H_2$ molecule (Table 4.8).

## BOND ORDER

The data of Tables 4.8 and 4.9 enable the definition of the bond order between two atoms, BO. Table 4.9 has shown that both the **bonding** and **anti-bonding** MOs (Figure 4.8), have a maximum capacity of two electrons, with opposite spins, $m_s \pm \frac{1}{2}$.

Bond Order (BO)
$$= \frac{\text{(No. of bonding electrons} - \text{Number of anti-bonding electrons)}}{2}.$$
Single bond, BO $= 1$; double bond, BO $= 2$; triple bond, BO $= 3$.

Thus the $\sigma^{b2}$ configuration represents a BO of 1, equivalent to a single bond, as in the $H_2$ diatomic molecule, *i.e.* a two-electron pair single bond. In the $H^+$ cation, with only a single electron in the $\sigma^{b1}$ bonding MO, this represents a BO of 0.5, as does the $He_2^+$ cation $\sigma^{b2}\sigma^{*1}$, as the third electron is in the **anti-bonding** molecular orbital which cancels the bonding effect of one of the electrons in the bonding MO, leaving a net single electron in a **bonding** MO and hence a BO of 0.5. In the $He_2$ molecule, $\sigma^{b2}\sigma^{*2}$, the two electrons in the anti-bonding MO cancel out the two electrons in the bonding MO, giving a BO of zero.

In the double bond of the $O_2$ molecule the BO order is 2.0, and in the triple bond of the $N_2$ molecule the BO is 3.0.

*Chapter 5*

# The Lewis Structures of Molecules, Cations and Anions, Including Oxyanions

### AIMS AND OBJECTIVES

This chapter demonstrates the use of the one electron orbital configuration of the elements to predict the Lewis structures of simple molecules, cations and anions, including oxyanions, as an introduction to the two electron pair bond. A systematic **Working Method** is suggested for determining the Lewis structure of a simple molecule, anion or cation, involving the one-electron orbital configuration of the central element, the number of terminal atoms, the contribution of the number of multiple bonds and the overall charge of the species. No attempt will be made to describe the shapes of these simple polyatomic species; this will be deferred until the next chapter on the use of the Valence State Electron Pair Repulsion Theory.

### INTRODUCTION

In the early 1900s Thompson, Bohr and Rutherford developed the early understanding of the electronic structure of the atom and its use to establish the structure of the Long Form of the Periodic Table (Figure 5.1). This now forms the basis of our understanding of the physical properties of the elements, *i.e.* their atomic number, atom size, ionisation energies and electron affinity or attachment enthalpy, and the chemical properties, *i.e.* characteristic or group oxidation number, oxidation number, variable valence, the stoichiometry of simple compounds and, ultimately, the stoichiometry of chemical reactions. This understanding quickly developed into the new Quantum Theory to describe the prop-

LONG FORM of the PERIODIC TABLE of the ELEMENTS

| 1 | 2 | 3 | 4 | 5 | 6 | 7 | 8 | 9 | 10 | 11 | 12 | 13 | 14 | 15 | 16 | 17 | 18 |
|---|---|---|---|---|---|---|---|---|---|---|---|---|---|---|---|---|---|
| IA | IIA | IIIB | IVB | VB | VIB | VIIB | VIIIB | | | IB | IIB | IIIA | IVA | VA | VIA | VIIA | VIIIA |
| H $1s^1$ | | | | | | | | | | | | | | | | | He $1s^2$ |
| Li $2s^1$ | Be $2s^2$ | | | | | | | | | | | B $2p^1$ | C $2p^2$ | N $2p^3$ | O $2p^4$ | F $2p^5$ | Ne $2p^6$ |
| Na $3s^1$ | Mg $3s^2$ | | | | | | | | | | | Al $3p^1$ | Si $3p^2$ | P $3p^3$ | S $3p^4$ | Cl $3p^5$ | Ar $3p^6$ |
| K $4s^1$ | Ca $4s^2$ | Sc $3d^1$ | Ti $3d^2$ | V $3d^3$ | Cr $3d^4$ | Mn $3d^5$ | Fe $3d^6$ | Co $3d^7$ | Ni $3d^8$ | Cu $3d^9$ | Zn $3d^{10}$ | | | | | | |

**Figure 5.1** *A shortened version of the Long Form of the Periodic Table – 30 elements*

| I | II | III | IV | V | VI | VII | VIII |
|---|----|-----|----|---|----|-----|------|
| 1 | 2 | 13 | 14 | 15 | 16 | 17 | 18 |
| H• | | | | | | | He: |
| $1s^1$ | | | | | | | $1s^2$ |
| Li• | Be: | B: | •C: | •N: | •O: | •F: | :Ne: |
| [He] $2s^1$ | $2s^2$ | $2s^2 2p^1$ | $2s^2 2p^2$ | $2s^2 2p^3$ | $2s^2 2p^4$ | $2s^2 2p^5$ | $2s^2 2p$ |
| Na• | Mg: | Al: | •Si: | •P: | •S: | •Cl: | :Ar: |
| [Ne] $3s^1$ | $3s^2$ | $3s^2 3p^1$ | $3s^2 3p^2$ | $3s^2 3p^3$ | $3s^2 3p^4$ | $3s^2 3p^5$ | $3s^2 3p$ |
| K• | Ca: | | | | | | |
| [Ar] $4s^1$ | $4s^2$ | | | | | | |

**Figure 5.2** *Abbreviated Periodic Table, valence shell configuration, dot form*

erties of *s*-, *p*-, *d*- and *f*-orbitals, and their use to consolidate the difference between ionic and covalent bonding, namely Pauling's Electronegativity values. Using the description of the electron pair bond, the shapes of simple covalent molecules, cations and anions could be understood, ideas that are hidden beneath the wealth of *X*-ray crystallographic data now available. In this period of rapid progress, the contribution that B. N. Lewis made in describing the valence shell configuration of the atoms (Figure 5.2), the closed valence shell configurations (inert gas cores) and the simple Lewis structures of covalent molecules is largely ignored.

This suggests that the present generation of students will be unable to make the connection between the electron configurations of the elements and the shapes of simple molecules, cations and anions, including the oxyanions, with any confidence. It also means that many students who are taught chemistry as a subsidiary subject to biology, botany, zoology, biochemistry or environmental chemistry do not appreciate the structure of the simple oxyacids of the main group elements, namely carbonates, nitrates, nitrites, silicates, phosphates and sulfates, a group of compounds that are so important in the above subjects. The primary objectives of this chapter are:

(a) To suggest and apply a logical Working Method to write the Lewis structures of simple Main Group covalent compounds.
(b) To extend this working method to the structures of the simple oxyacids and oxyanions of the Main Group elements.

With these objectives, it is hoped that the importance of Lewis structures to the teaching of elementary chemistry is reiterated and that this chapter will emphasise their value as a connecting bridge between the electron configuration of the elements and the shape of the simple Main Group compounds, cations and anions, including the oxyacids, as described in Chapter 6.

## THE WORKING METHOD FOR DRAWING LEWIS STRUCTURES

The Working Method for drawing Lewis structures for molecules, anions and cations involves the following steps:

1. Characterise the species given as a neutral molecule, cation or anion and identify the central atom.
2. Draw a ball and stick diagram for the polyatomic molecules, anions or cations, including localised double bonds.
3. Determine the total number of valence shell electrons present, add electrons for a negatively charged anion, subtract electrons for a positively charged cation.
4. Assign two electrons for all single bonds as **xx** (to identify the single bonded electron pairs).
5. Distribute the remaining electrons not involved in bonding, as lone pairs, **oo** (to distinguish the lone pairs of electrons from the bonding pairs of electrons) to complete the octets, where possible, but excluding H which only requires two electrons.
6. Where possible complete octets by converting single bonds to double bonds, especially for terminal oxygen atoms with a free valence.
7. Finally, redraw all electrons as one type (● ●) to remove the distinction between the above origins of the electrons and check the total number of electrons, with that determined in (3) above.

This Working Method will be applied systematically to a number of examples, in pairs, of molecules, anions and cations, including oxy-anions, to illustrate its application.

### Example 1: Methane ($CH_4$) and Carbon Tetrachloride ($CCl_4$)

Methane and carbon tetrachloride represent the simplest type of covalent polyatomic molecule, involving an $s^2p^6$ valence shell configuration of eight electrons, four shared electron pairs and four covalent bonds and only differing in the terminal atoms. H has a shared $s^2$ configuration and Cl has an $s^2p^6$ configuration, with only one of its four electron pairs shared with the carbon atom.

(1) Identify a neutral molecule with main group central atom.
(2) Draw a ball and stick diagram.

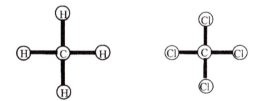

(3) Determine the total number of valence shell electrons.

$$s^2p^2 \quad s^1 \qquad\qquad s^2p^2 \quad s^2p^5$$
$$C(4) + 4H(1) = 8 \qquad C(4) + 4Cl(7) = 32$$

(4) Assign two electrons for all single bonds as **xx** pairs, to identify the single bond electron pairs.

(5) Distribute the remaining electrons as **oo** pairs (simply to identify the lone pairs of electrons) to complete the octets. As H only requires two electrons, $CH_4$ is already complete in step (4), but each Cl in $CCl_4$ requires three additional electron pairs.

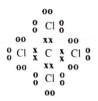

(6) This step is not required as there are no double bonds.

(7) Finally, redraw all electron pairs as •• pairs, to remove the distinction between the origins of the electrons.

This completes the Lewis structures of $CH_4$ and $CCl_4$ in which each C has an octet with four bonding pairs, each H has an $s^2$ configuration involving a shared electron pair, and each of the four Cl atoms has an octet, involving one shared electron pair (bonding) and three unshared electron pairs (long pairs).

### Example 2: The Ammonium Cation ($NH_4^+$) and the Tetrafluoroborate Anion ($BF_4^-$)

The Working Method applies to both the cation and the anion in $NH_4BF_4$.

(1) Identify the charges on the cation and the anion: $NH_4^+$ and $BF_4^-$.

(2) Draw a ball and stick diagram of the polyatomic $NH_4^+$ cation and of the $BF_4^-$ anion.

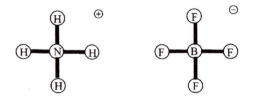

(3) Determine the total number of valence shell electrons in each cation and anion.

$s^2p^3$  $s^1$      positive      $s^2p^1$  $s^2p^5$      negative

$N(5) + 4H(1) - 1 = 8$          $B(3) + 4F(7) + 1 = 32$

(4) Assign two electrons for all single bonds as **xx** pairs, to identify the bonding pairs.

(5) Distribute the remaining electrons as **oo** pairs (to identify lone pairs of electrons). As H only requires two electrons, the Lewis structure of $NH_4^+$ is complete, total $= 8$, but the Fs of $BF_4^-$ require three further electron pairs to complete their octets, total $= 32$.

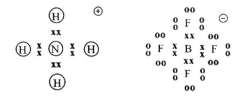

(6) As there are no double bonds in this step is not required.
(7) Finally, redraw all electrons as •• pairs, to remove the distinction between the origins of the electrons.

This completes the determination of the Lewis structures of the cation, $NH_4^+$, and the anion, $BF_4^-$. In the former, nitrogen has an octet with four shared bonding electron pairs, with each H atom having an $s^2$ configuration, involving a shared electron pair. In the latter, the boron atom has an octet involving four shared electron pairs (bonding), and each F atom has an octet involving one shared electron pair (bonding), and three unshared electron pairs (lone pairs).

### Example 3: Ammonia (NH₃) and Water (OH₂)

The Working Method also applies to molecules with an octet of electrons on the central atom, but with one or more of the four electron pairs present as lone pairs, as in the case of $NH_3$ and $OH_2$.

(1) Both molecules are identified as neutral molecules.
(2) Draw a ball and stick diagram.

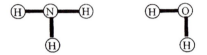

(3) Determine the total number of valence shell electrons in each molecule.

$$s^2p^3 \quad s^1 \qquad\qquad s^2p^4 \quad s^1$$
$$N(5) + 3H(1) = 8 \qquad O(6) + 2H(1) = 8$$

(4) Assign two electrons for all single bonds as **xx** pairs, to identify the bonding electron pairs.

$$\text{(H)} \; {\overset{x}{\underset{x}{}}} \; \text{(N)} \; {\overset{x}{\underset{x}{}}} \; \text{(H)} \qquad\qquad \text{(H)} \; {\overset{x}{\underset{x}{}}} \; \text{(O)}$$
$$\mathbf{xx} \qquad\qquad\qquad \mathbf{xx}$$
$$\text{(H)} \qquad\qquad\qquad \text{(H)}$$

(5) Distribute the remaining electrons as **oo** pairs to identify lone pairs of electrons). As hydrogen only requires two electrons, only one **oo** electron pair is added to the nitrogen atom, and two **oo** electron pairs to the oxygen atom. This completes the octet, in both cases, of 8 electrons.

$$\mathbf{oo} \qquad\qquad\qquad\qquad \mathbf{oo}$$
$$\text{(H)} \; {\overset{x}{\underset{x}{}}} \; \text{(N)} \; {\overset{x}{\underset{x}{}}} \; \text{(H)} \qquad\qquad \text{(H)} \; {\overset{x}{\underset{x}{}}} \; \text{(O)} \; {\overset{o}{\underset{o}{}}}$$
$$\mathbf{xx} \qquad\qquad\qquad\qquad \mathbf{xx}$$
$$\text{(H)} \qquad\qquad\qquad\qquad \text{(H)}$$

(6) As there are no double bonds in $NH_3$ or $OH_2$, this step is not required.
(7) Finally redraw all electrons as •• pairs, to remove the distinction between the origins of the electrons.

This completes the determination of the Lewis structures of $NH_3$ and $OH_2$. Both the N and O atoms have closed octets, but in $NH_3$ this involves three bonding pairs, and one lone pair, and in $OH_2$ this involves two bonding pairs and two lone pairs of electrons.

### Example 4: Beryllium Dihydride ($BeH_2$) and Boron Trifluoride ($BF_3$)

Not all molecules involves Lewis structures with complete octets of electrons: for example $BeH_2$ and $BF_3$ involve four and six electron shells, respectively, a configuration that makes them reactive towards electron donors. The Working Method still applies to such electron deficient systems.

(1) Both $BeH_2$ and $BF_3$ can be identified as neutral molecules.
(2) Draw a ball and stick diagram.

(3) Determine the total number of valence shell electrons.

$$s^2 \quad\quad s^1 \qquad\qquad s^2p^1 \quad s^2p^5$$
$$Be(2) + 2H(1) = 4 \quad\quad B(3) + 3F(7) = 24$$

(4) Assign two electrons for all single bonds as **xx** pairs.

(5) Distribute the remaining electrons from the total as **oo** pairs, to complete the octets.

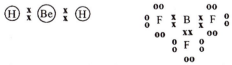

(6) This step is not required as there are no double bonds.

(7) Finally redraw all electrons as •• pairs, to remove the distinction between the origins of the electrons.

Check that the total number of valence electrons agrees with that determined in (3) above, 4 and 24 respectively. This completes the Lewis structures of $BeH_2$ and $BF_3$. The former has only two electron pairs, both shared, whereas the latter has three shared electron pairs, a total of four and six electrons rather than the eight required for the octet.

### Example 5: Phosphorus Pentachloride ($PCl_5$) and Sulfur Hexafluoride($SF_6$)

In contrast, $PCl_5$ and $SF_6$ involve valence shells greater than an octet, but the Working Method still applies to these molecules containing expanded octets.

(1) Both are neutral molecules.
(2) Draw a ball and stick diagram.

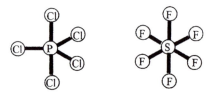

(3) Determine the total number of valence shell electrons.

$$s^2p^3 \quad s^2p^5 \qquad s^2p^4 \quad s^2p^5$$
$$P(5) + 5Cl(7) = 40 \qquad S(6) + 6F(7) = 48$$

(4) Assign two electrons for all single bons as xx pairs.

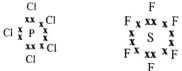

(5) Distribute the remaining electrons as oo pairs, to complete the octets.

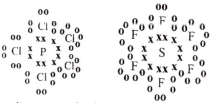

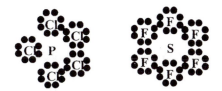

(6) This step is not required as there are no double bonds.

(7) Finally, redraw all electrons as •• pairs, to remove the distinction between the origins of the electrons.

This completes the Lewis structures of $PCl_5$ and $SF_6$ respectively, in which the P has a ten electron core involving five shared electron pair bonds, and S has a twelve electron core involving six shared electron pair bonds. Neither the central P or S atoms has a lone pair of electrons.

### Example 6: 1,1-Dichloromethanone ($Cl_2CO$) and Ethene ($C_2H_4$)

The Working Method can also be applied to molecules containing multiple bonds, as in the structures of $Cl_2C=O$ and $H_2C=CH_2$.
**Note:** the molecule of ethene contains two separate, but equivalent, C atom centres, which are treated separately.

(1) Both structures are identified as neutral molecules.
(2) Draw a ball and stick diagram.

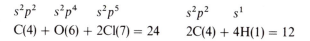

(3) Determine the total number of valence shell electrons.

$$s^2p^2 \quad s^2p^4 \quad s^2p^5 \qquad s^2p^2 \quad s^1$$
$$C(4) + O(6) + 2Cl(7) = 24 \qquad 2C(4) + 4H(1) = 12$$

(4) Assign two electrons for all single bonds as **xx** pairs, including the single bond involved in the double bond.

**Note:** this only provides a six electron shell for the C and O atoms involved in the double bonds.

(5) Distribute the remaining electrons as **oo** pairs, to complete the octets.

(6) In the case of $Cl_2CO$ this gives a total of 24 electrons, giving the two Cl atoms and the O atom a complete octet (two shared pairs and three lone pairs), but leaves the carbon with only a shared sextet. The C atom is thus electron deficient, and the O atom relatively electron rich. The latter is only involved in one single bond, whereas it normally prefers two single bonds, as in $OH_2$. It is this additional single bond capacity that is described as a **free valence** on the O atom. In $Cl_2CO$, the initial 'free valence' is linked into the localised double bond between the C and O atoms, by the terminal oxygen atom donating a share of one of its lone pairs to the carbon atom, thus retaining its octet (two shared electron pairs plus two lone pairs), and increases the C sextet to an octet of four shared pairs, two single bond pairs and two double bond pairs.

In the case of ethene only a single additional **oo** pair is involved, as both the C atoms are equivalent, with electron deficient sextets. The two electrons could be added singly to each C atom, but this only gives each carbon atom a total of seven electrons, still an incomplete octet. By sharing the additional **oo** pair of electrons equally between the two C atoms, this gives a double bond between the C atoms and completes the octet of both C atoms.

(7) Finally, redraw all the electrons as equal electron pairs **●●** to remove the distinction between the origin of the electrons.

This completes the Lewis structures of these two double bonded molecules $Cl_2CO$ and $H_2CCH_2$. In $Cl_2CO$, all four atoms have complete octets: the two Cl atoms involve one shared and three lone pairs, the O atom two shared and two lone pairs, and the C atom four shared pairs of electrons. On both the oxygen and carbon atoms, two of the shared electron pairs are involved in a double bond. In $H_2CCH_2$, each of the Four H atoms has a single shared two-electron pair, and the two C atoms involve four shared electron pairs, two of which are involved in a double bond.

### Example 7: Ethyne ($C_2H_2$)

The Working Method may also be applied in the case of a molecule containing a triple bond, such as ethyne, HCCH.

(1) Ethyne is a neutral molecule, containing two carbon atoms that are treated separately.

(2) Draw a ball and stick diagram.

(3) Determine the total number of electrons.

$$s^1 \qquad s^2p^2$$
$$2H(1) + 2C(4) = 10$$

(4) Assign two electrons for all single bonds as **xx** pairs, including that involved in the triple bond.

$$H \underset{x}{\overset{x}{\phantom{|}}} C \underset{x}{\overset{x}{\phantom{|}}} C \underset{x}{\overset{x}{\phantom{|}}} H$$

**Note:** this only provides a four electron shell for each C atom involved in the triple bond.

(5) Distribute the remaining electrons as **oo** pairs.

$$H \underset{x}{\overset{x}{\phantom{|}}} C \overset{o\,o}{\underset{o\,o}{\phantom{|}}} C \underset{x}{\overset{x}{\phantom{|}}} H$$

(6) As there are only two electron pairs, these could be assigned separately to each C atom, but these would not attain an octet,

only an incomplete sextet for each C atom. By sharing both additional **oo** electron pairs between the two C atoms, not only is a triple bond formed, but both C atoms acquire a complete shared octet.

(7) Finally, redraw all the electrons as •• pairs, to remove the distinction between the origins of the electrons.

$$H \overset{\bullet}{\underset{\bullet}{\bullet}} C \overset{\bullet}{\underset{\bullet}{\bullet}} C \overset{\bullet}{\underset{\bullet}{\bullet}} H$$

This completes the Lewis structure of ethyne $C_2H_2$ with a triple bond. In $C_2H_2$ both C atoms have complete octets involving four shared electron pairs, one with the H atom and three with the second C atom. Each H atom has a single shared electron pair.

## THE OXYACIDS AND OXYANIONS OF THE MAIN GROUP ELEMENTS

The ideas of the valence shell configuration, oxidation number, Lewis structure and the shape of Main Group compounds are brought together in discussing a very important group of compounds of the Main Group elements, namely, the oxyacids and their oxyanions (Table 5.1).

Before the Working Method can be applied to the oxyacids and oxyanions of the Main Group elements a number of structural features in these compounds need to be resolved.

### The Position of the Hydrogen Atoms in the Oxyacids

From the formal oxidation numbers, all central elements in the oxyacids involve a positive oxidation number (III to VII), all the oxygen atoms involve a negative oxidation number ( − II) and all the hydrogen atoms involve a positive oxidation number (I), all with formal inert gas cores. In view of the positive oxidation state of all the central Main Group elements of Table 5.1 and the positive oxidation state of the H atoms in the acids, the latter are always associated with the electronegative oxygen atoms and **not** with the central Main Group elements. Thus the ball and stick diagrams of $H_3BO_3$ and $H_4SiO_4$ may be represented as:

**Table 5.1** *Some oxyacids and oxyanions of the main group elements*

| $s^2p^1$ | $s^2p^2$ | $s^2p^3$ | $s^2p^4$ | $s^2p^5$ |
|---|---|---|---|---|
| III | IV | V | VI | VII |
| $H_3BO_3$ | $H_2CO_3$ | $HNO_3$ | | |
| $BO_3^{3-}$ | $CO_3^{2-}$ | $NO_3^-$ | | |
| | $H_4SiO_4$ | $H_3PO_4$ | $H_2SO_4$ | $HClO_4$ |
| | $SiO_4^{4-}$ | $PO_4^{3-}$ | $SO_4^{2-}$ | $ClO_4^-$ |

This association of the hydrogen atoms of these Main Group elements with the electronegative terminal oxygen atoms, to retain their inert gas cores, suggests that these acids should be written as $B(OH)_3$ and $Si(OH)_4$ rather than the traditional $H_3BO_3$ and $H_4SiO_4$ formulations.

Equally, the corresponding oxyanions may be represented as involving a localised negative charge associated with each oxygen atom from which a single proton has been removed.

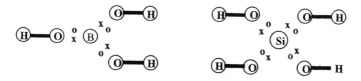

The corresponding Lewis structures may be represented as:

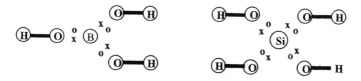

All the O and Si atoms have a [Ne] inert gas core, but the B atom has a non-inert gas core of $s^2p^4$ and the H atom has a [He] inert gas core. In the corresponding oxyanions, the removal of the H atoms as $H^+$ leaves a negative charge associated with each of the terminal oxygen atoms from which a proton, $H^+$, has been removed and maintains the [Ne] inert gas core configuration of these terminal O atoms. Conversely, in writing the ball and stick structure of any of the oxyanions of Table 5.1, the negative charges of the oxyanions are specifically localised on the terminal oxygen atoms and are **not** associated with the central Main Group element. This notation is particularly important in writing the Lewis structures of the oxyanions of Table 5.1 and in determining the shape of the oxyanions by VSEPR theory, as illustrated in Chapter 6.0.

### The Free Valence of the Terminal Oxygen Atoms

However, this still leaves a problem with the structures of the remaining oxyacids of Table 5.2. Each oxygen atom associated with a hydrogen atom is divalent, one electron pair bond to hydrogen and one to the central element. The remaining O atoms only involve one single bond to the central atom; such terminal oxygen atoms have an unused bonding potential or **free valence**. In the case of $H_2CO_3$ and $H_2SO_4$, two H atoms satisfy the divalency of two of the terminal O atoms, but the remaining O atoms are only singly bound to the central atom and are left with a free valence, one and two, for $H_2CO_3$ and $H_4SO_4$, respectively.

In the case of $H_2CO_3$, the single oxygen atom with a 'free valence' is associated with a carbon atom with an incomplete octet on the carbon atom, four electrons from C and three electrons from three separate oxygen atoms, both of which can be resolved by the formation of a double bond between the central carbon atom and the terminal oxygen atom with the 'free valence':

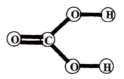

The involvement of a single C–O double bond in $H_2CO_3$ can also be extended to the carbonate oxyanion, $CO_3^{2-}$.

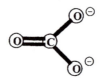

The two free valences of $H_2SO_4$ can be resolved in a similar way, for both the acid and the free anion.

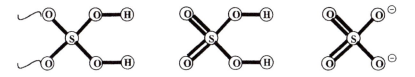

However, in this case, although the neon inert gas configurations of the terminal oxygen atoms are maintained, that of the central S atom needs to be expanded to twelve electrons, *i.e.* six shared electron pairs. This type of octet expansion has been seen previously in $SF_6$.

In the same way the ball and stick diagrams of the remaining oxyacids and oxyanions may be considered, with appropriate expansion of the central element octet.

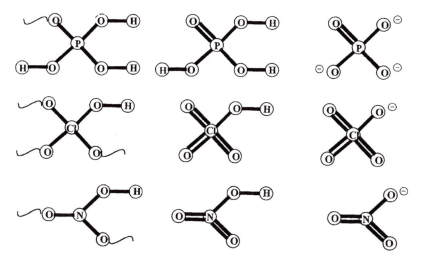

Equally, the oxyacids in lower oxidation states than the group oxidation state may be treated, again with appropriate expansion of the central element octet.

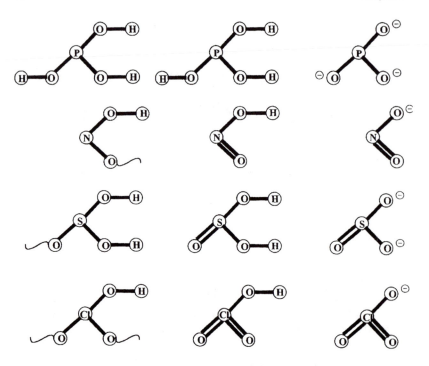

**Resonance in the Structures of the Oxyanions**

The problem associated with the presence of **free valences** in representing the structural formula of the oxyanions of the Main Group elements was resolved by converting these to localised double bonds. However, this implies that in, for example, the $CO_3^{2-}$ oxyanion there are two types of C–O bonds, one double bond and two single bonds, with the former slightly shorter than the latter. In practice, $X$-ray crystallography shows no significant difference between the C–O distances in the free $CO_3^{2-}$ anion, or in any of the element–O distances in the free oxyanions described above. This suggests that the actual structure of the $CO_3^{2-}$ oxyanion is best represented as a **resonance hybrid** of the following three structures:

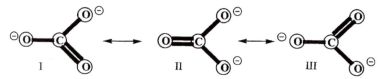

A corresponding set of equivalent structures is involved with the parent acid, $H_2CO_3$:

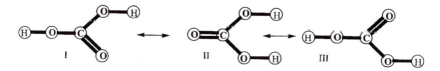

**Resonance:** In these three extreme resonance structures of the carbonate oxyanion, $CO_3^{2-}$, normal single and double bonds are present, but neither of them separately represents the C–O bonding in the $CO_3^{2-}$ anion. In a resonance hybrid structure (*i.e.* the three structures I, II and III in the case of the $CO_3^{2-}$ anion) the three individual canonical structures do not separately exist, but the actual structure involves an equal contribution from the three extreme canonical structures, namely, $\frac{1}{3}I + \frac{1}{3}II + \frac{1}{3}III$, thus giving three equal C–O bonds of $1\sigma + \frac{1}{3}\pi$ bond character. This is then consistent with the $X$-ray crystallography view of the bonding of the $CO_3^{2-}$ anion, that it is trigonal planar (Chapter 6), with three equivalent C–O distances of 120 pm, reflecting three equal types of C–O bonding, namely $1\sigma + \frac{1}{3}\pi$. This type of double bond character is best described by Molecular Orbital theory. The actual structure involves not three localised double bonds, but a delocalised double bond between the three CO distances, so that each CO distance has one-third double bond character. However this does not invalidate the use of localised double bonds to determine both the Lewis structures and the shapes by VSEPR theory of the oxyacids and oxyanions of the Main Group elements.

## THE APPLICATION OF THE WORKING METHOD TO THE LEWIS STRUCTURES OF OXYANIONS

### Example 1: Carbonic Acid, $H_2CO_3$

(1) Carbonic acid is a neutral diprotic acid:
$$H_2CO_3 \rightarrow 2H^+ + CO_3^{2-}$$

(2) Draw a ball and stick diagram. Associate the two H atoms with two of the terminal O atoms to form O to H single bonds, leaving the third O atom with only one covalent bond, namely a free valence, and convert it to a localised double bond.

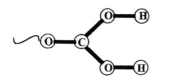   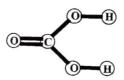

(3) Determine the total number of valence shell electrons.

$$s^1 \qquad s^2p^2 \quad s^2p^4$$
$$2H(1) + C(4) + 3O(6) = 24$$

(4) Assign two electrons for all single bonds as **xx** pairs.

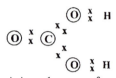

(5) Distribute the remaining electrons from the total as **oo** pairs, to complete the octets of the oxygen atoms.

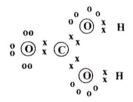

This leaves all the O atoms with octets, the H atom with a shared electron pair, the C atom with only a six electron shell and one of the O atoms with a free valence.

(6) Redistribute one lone pair of electrons on the non-hydrogen linked O atom, *i.e.* the one with the free valence, to be shared with the central C atom, to form a C–O localised double bond.

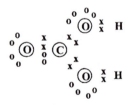

(7) Finally, redraw all electron pairs as ●● pairs, to remove the distinction between the origin of the electrons and determine that the total number of electrons agrees with the total in (3).

This completes the Lewis structure of $H_2CO_3$, in which the central C atom and three O atoms have a [Ne] inert gas octet and the H atom a [He] core. Two of the O atoms have two lone pairs and share single electron pairs with a H atom and the central C atom. The third O atom has two lone pairs and shares two electron pairs with the central C atom to form a double bond. The central C atom shares two electron pairs with two separate O atoms and two further electron pairs with the third O atom to form a double bond. The two H atoms each share a pair of electrons with two separate O atoms to form single bonds.

As suggested above the actual structure of $H_2CO_3$ is best represented as three equivalent resonance structures, rather than with a single localised double bond, thus:

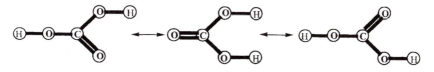

The Working Method may also be used to determine the Lewis structure of the $CO_3^{2-}$ oxyanion, with exactly the same results as above, except that the two H atoms are replaced by two negative charges, thus:

A comparable resonance hybrid structure can also be drawn for the $CO_3^{2-}$ anion:

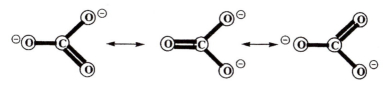

## Example 2: Sulfuric Acid ($H_2SO_4$)

(1) Sulfuric acid is a neutral diprotic acid: $H_2SO_4 \rightarrow 2H^+ + SO_4^{2-}$.
(2) Draw a ball and stick diagram. Associate the two H atoms with two of the terminal O atoms to form O to H single bonds, leaving two O atoms with only one covalent bond, namely a free valence.

(3) Determine the total number of valence shell electrons.

$$s^1 \qquad s^2p^4 \quad s^2p^4$$

$$2H(1) + S(6) + 4O(6) = 32$$

(4) Assign two electrons for all single bonds as **xx** pairs.

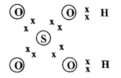

(5) Distribute the remaining electrons from the total in (3) as **oo** pairs, to complete the octets of the oxygen atoms.

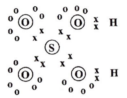

This leaves all the O atoms with octets, the H atom with a shared electron pair, the S atom with an octet, and two of the O atoms with free valences.

(6) Redistribute one lone pair of electrons on each of the non-hydrogen linked O atoms, *i.e.* the two with the free valence, to be shared with the central S atom, to form S–O localised double bonds.
**Note:** this increases the number of shared electron pairs on the S atom from four to six electron pairs and the valence shell from eight to twelve electrons, as in $SF_6$ above, namely an expanded octet.

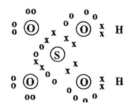

(7) Finally, redraw all electron pairs as •• pairs, to remove the distinction due to the origin of the electrons and check that the total number of electrons agrees with the total in (3) above.

This completes the Lewis structure of $H_2SO_4$, in which the central S atom has an expanded octet of twelve electrons, the four O atoms have a [Ne] inert gas core octet and the H atoms a [He] core. Two of the O atoms have two lone pairs and share single electron pairs with a H atom and the central S atom. The remaining two O atoms involve two lone pairs and share two electron pairs with the central S atom to form two double bonds. The central S atom shares four electron pairs with four separate O atoms and two further electron pairs with two of these O atoms to form two localised double bonds. The two H atoms each share a pair of electrons with two separate O atoms to form single bonds.

As suggested above the actual structure of $H_2SO_4$ is best represented as a number of equivalent structures, rather than with single localised double bonds. Three of the eight possible equivalent structures are shown thus:

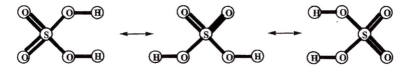

The Working Method may also be used to determine the Lewis structure of the $SO_4^{2-}$ oxyanion, with exactly the same results as above, except that the two H atoms are replaced by two negative charges, thus:

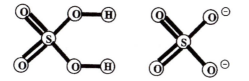

A comparable set of equivalent resonance structures can also be drawn for the oxyanion, as for the oxyacid:

## THE USE OF FORMAL CHARGES

A provisional Lewis structure may contain the correct bonding framework, but the distribution of the valence electrons may not be the one that gives the maximum stability. The correct stereochemistry is predicted by the valence shell configuration using VSEPR theory, as shown in Chapter 6. A concept called **formal charge** (FC) can be used to predict which structure of a number of alternative structures is the most reasonable for a Lewis structure. The **formal charge** (FC) on any atom in a Lewis structure can be defined as:

FC = Number of valence electrons of the free atom
     − Number of valence electrons in the Lewis structure.

Using the following rules:

1. Half the electrons in a bond are assigned to each atom in the bond.
2. Both electrons of an unshared pair are assigned to an atom.

A few simple examples demonstrate the application of the calculation of FC for each of the separate atoms:

*Formal charge*

|  |  | Formal Charge |
|---|---|---|
| $\begin{array}{cc} \text{oo} & \text{oo} \\ {}^{\text{o}}_{\text{o}}\text{F}{}^{\text{o}}_{\text{o}}\text{F}{}^{\text{o}}_{\text{o}} \\ \text{oo} & \text{oo} \end{array}$ | F [He] $2s^2 2p^5$ | FC = 7 - 7 = 0 |
| $\begin{array}{c} \text{oo} \\ {}^{\text{o}}_{\text{o}}\text{F}{}^{\text{o}}_{\text{o}}\text{H} \\ \text{oo} \end{array}$ | H $1s^1$ | FC = 1 - 1 = 0 |
|  | F [He] $2s^2 2p^5$ | FC = 7 - 7 = 0 |
| $\begin{array}{c} \text{oo} \\ \text{(H)}\ {}^{x}_{x}\text{(N)}\ {}^{x}_{x}\ \text{(H)} \\ \text{xx} \\ \text{(H)} \end{array}$ | N [He] $2s^2 2p^3$ | FC = 5 - 5 = 0 |

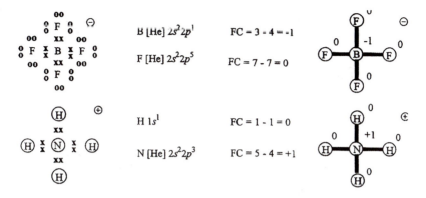

B [He] $2s^2 2p^1$    FC = 3 - 4 = -1

F [He] $2s^2 2p^5$    FC = 7 - 7 = 0

H $1s^1$    FC = 1 - 1 = 0

N [He] $2s^2 2p^3$    FC = 5 - 4 = +1

These examples illustrate the application of FC to simple examples which can then be extended to double and tripe bonds:

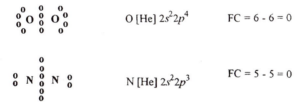

O [He] $2s^2 2p^4$    FC = 6 - 6 = 0

N [He] $2s^2 2p^3$    FC = 5 - 5 = 0

In certain cases, the FC of the alternative Lewis structures can help to identify the preferred structure. Thus the Lewis structure for $BF_3$ has two alternative structures:

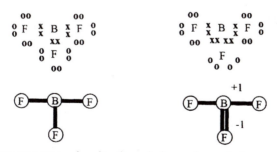

In the former structure, despite there being a formal charge of zero on both the boron and the fluorine, the B atom only has a valence shell of six rather than that of the expected eight electrons. In the alternative double bonded structure, although this provides a shared octet for the B atom, the formal charge difference of $-1 - (1) = -2$ is too large to represent the most stable structure; thus the three single bonded structure is preferred.

In the case of $H_2SO_4$, two alternative structures are possible:

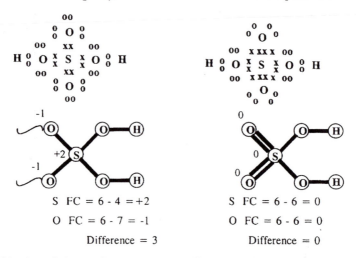

S  FC = 6 - 4 = +2          S  FC = 6 - 6 = 0

O  FC = 6 - 7 = -1          O  FC = 6 - 6 = 0

Difference = 3             Difference = 0

In this the minimum formal charge difference is for the double bonded structure, which is therefore preferred.

As normally written nitric acid involves two N–O double bonds, in which the N atom has an octet expanded from eight to ten, which is unusual for a first short period element. However, the alternative of one $N=O$ and one $N^+-O^-$ bond involves a difference of 2 in the formal charge of the latter, and consequently the former structure is preferred.

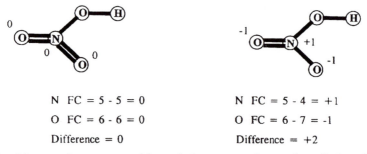

N  FC = 5 - 5 = 0            N  FC = 5 - 4 = +1

O  FC = 6 - 6 = 0            O  FC = 6 - 7 = -1

Difference = 0             Difference = +2

In this way the concept of formal charge may be used to distinguish the most stable structure of alternative Lewis structures of simple inorganic molecules, cations, and anions, including oxyacids and oxyanions.

## SUMMARY

Used in this way the Lewis structures of simple molecules, anions and cations of the Main Group elements, including acids and oxyanions, form an effective bridge between the valence shell electron configuration of an element, $s^m p^n$, determined from its position in the Periodic Table and the description of the bonding of the compound.

# Chapter 6

# Shape and Hybridisation

## AIMS AND OBJECTIVES

Previous chapters have shown how the electron configurations of the elements determine the long form of the Periodic Table (Figure 2.11), their oxidation numbers and variable valence, the stoichiometry factors in determining reactions in volumetric chemistry and the types of bonding in molecules, *i.e.* ionic or covalent. Chapter 5 has shown how the electron configuration of the elements may be used to determine the Lewis structures of simple molecules, anions and cations, including oxyanions. This chapter extends this use of the electron configuration of the elements to the prediction of the shapes of simple covalent molecules, cations and anions, including oxyacids and oxyanions, using the Valence State Electron Pair Repulsion (VSEPR) theory. A systematic numerical Working Method is suggested to determine the basic shapes of simple covalent molecules, anions and cations by VSEPR theory, taking into account the electron configuration of the central element, the number of $\sigma$ bonds, the number of $\pi$ bonds and the presence of an overall positive or negative charge. Given the basic shape of a species the presence of lone pairs may then be used to predict the distortions from a regular stereochemistry, and the type of bond hybridisation involved can be suggested.

## THE SHAPES OF COVALENT MOLECULES

An important property of covalent molecules is that these covalent bonds have directional properties, and the molecules have three-dimensional shape. What determines this shape is the *number* of electron pair bonds in the *valence shell configuration* of the central atom (Figure 6.1), about which the shape of the molecule is described. The VSEPR theory

| Number of Electron pairs | Shape | Figure | Angles / ° |
|---|---|---|---|
| 2 | linear | | 180 |
| 3 | trigonal planar | | 120 |
| 4 | tetrahedral | | 109.5 |
| 5 | trigonal bipyramidal | | 90, 120, 180 |
| 6 | octahedral | | 90, 180 |

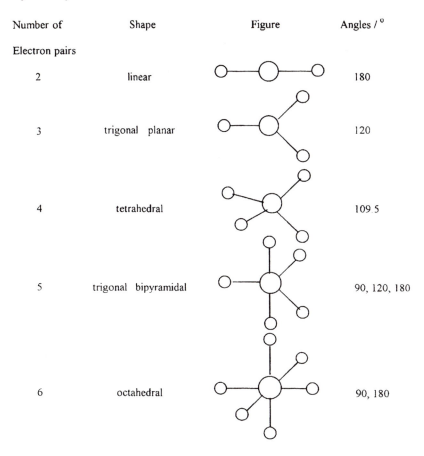

**Figure 6.1** *The predicted shapes of simple molecules, anions and cations*

describes a simple method to determine the shape of covalent molecules, anions and cations. It suggests that the arrangements in space about a central atom of the covalent bonds in a polyatomic molecule, cation or anion is primarily determined by the number of electron pairs in the valence shell of the central atom involved. It points out that the pairs of electrons would arrange themselves about the central atom to be as far apart as possible from each other and thus reduce their mutual repulsion. Thus, two electron pairs would arrange themselves linearly, *etc.* as set out in Table 6.1a, and illustrated in Figure 6.1. These simple ideas are readily extended to include the effect of lone pairs of electrons, E (Table 6.1b and Figure 6.2), and to the inclusion of localised double bond electrons, D (Table 6.1c and Figure 6.3).

**Table 6.1** *The shapes of simple molecules of formula* $AB_n$

(a) *Arrangement of electron pairs in molecules*

| Total number of electron pairs | Number of lone pairs | Formula | Shape | Angles/° | Examples |
|---|---|---|---|---|---|
| 2 | 0 | $AB_2$ | linear | 180 | $BeCl_2$, $CO_2$ |
| 3 | 0 | $AB_3$ | trigonal planar | 120 | $BF_3$ |
| 4 | 0 | $AB_4$ | tetrahedral | 109.5 | $CH_4$, $NH_4^+$, $BF_4^-$ |
| 5 | 0 | $AB_5$ | trigonal bipyramidal | 90, 120, 180 | $PCl_5$ |
| 6 | 0 | $AB_6$ | octahedral | 90, 180 | $SF_6$, $PCl_6^-$ |

(b) *Extended to include lone pairs (E) and bonded pairs of electrons (Figure 6.2)*

| Total number of electron pairs | Number of lone pairs | Formula | Shape | Angles/° | Examples | Experimental angle/° |
|---|---|---|---|---|---|---|
| 3 | 1 | $AB_2E$ | trigonal planar | 120 | $SnCl_2$ | 95 |
| 4 | 1 | $AB_3E$ | pyramidal | 109.5 | $NH_3$ | 107.4 |
| 4 | 2 | $AB_2E_2$ | bent | 109.5 | $OH_2$ | 104.5 |
| 5 | 1 | $AB_4E$ | trigonal pyramidal | 90, 120, 180 | $SF_4$ | 101.5, 173.1 |
| 5 | 2 | $AB_3E_2$ | T-shaped | 90, 180 | $ClF_3$ | 87.5, 175 |
| 5 | 3 | $AB_2E_3$ | linear | 180 | $XeCl_2$ | 180 |
| 6 | 1 | $AB_5E$ | square pyramidal | 90, 180 | $IF_5$ | 88.5, 161.8 |
| 6 | 2 | $AB_4E_2$ | square coplanar | 90, 180 | $XeF_4$ | 90, 180 |

(c) *Extended to include localised double bonds (D) (Figure 6.3)*

| Total number of electron pairs | Number of lone pairs | Number of double bonds | Formula | Shape | Angles/° | Examples | Experimental angles/° |
|---|---|---|---|---|---|---|---|
| 2 | 0 | 1 | $AB_2D$ | linear | 180 | — | — |
| 2 | 0 | 2 | $AB_2D_2$ | linear | 180 | $CO_2$ | 180.0 |
| 3 | 1 | 1 | $AB_2ED$ | bent | 120 | $NO_2^{-}$ | 115.0 |
| 3 | 1 | 2 | $AB_2ED_2$ | bent | 120 | — | — |
| 3 | 0 | 1 | $AB_3D$ | triangular planar | 120 | $CO_3^{2-}$ | 120.0 |
| 3 | 0 | 2 | $AB_3D_2$ | triangular planar | 120 | $NO_3^{-}$ | 120.0 |
| 4 | 1 | 1 | $AB_3ED$ | trigonal pyramidal | 109.5 | $SO_3^{2-}$ | 103.0 |
| 4 | 1 | 2 | $AB_3ED_2$ | trigonal pyramidal | 109.5 | $ClO_3^{-}$ | 106.0 |
| 4 | 0 | 1 | $AB_4D$ | tetrahedral | 109.5 | $(CH_3)_2P{=}O$ | 111.0 |
| 4 | 0 | 2 | $AB_4D_2$ | tetrahedral | 109.5 | $SO_4^{2-}$ | 109.0 |
| 4 | 0 | 3 | $AB_4D_3$ | tetrahedral | 109.5 | $ClO_4^{-}$ | 109.0 |

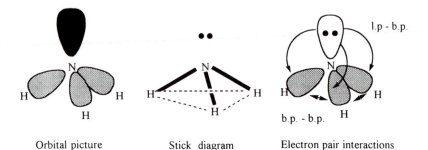

| Orbital picture | Stick diagram | Electron pair interactions |

**Figure 6.2** *Electron pairs, regions occupied by bonding and lone pairs*

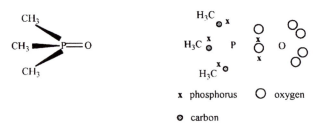

x phosphorus    O oxygen

O carbon

**Figure 6.3** *The electronic structure of the localized double bond in* $(CH_3)_3P = O$

## THE WORKING METHOD FOR USING VSEPR THEORY

In predicting the shape of any covalent species, it is strongly recommended that the student follows a given Working Method (see below). Working through the series of steps helps to avoid making simple mistakes and shows how easily one can deduce the shape of the molecule or ion. Nevertheless, the most important step in getting started is representing the molecule, anion or cation as a simple ball-and-stick model, in which the angles do not matter, but the number of covalent bonds is essential. Each bond to a terminal atom such as H in $CH_4$ or $NH_3$ is represented by a two-electron pair stick; each bond to a terminal halogen atom is represented by a two-electron pair stick, as both H and the halogens can be satisfied by a single covalent bond. If the terminal atom is oxygen, this cannot be satisfied by a single (stick) bond as it has a valence of two and thus, each terminal oxygen atom must be represented either as:

(a) by a double stick, representing a double bond to the central atom, (=O), or;
(b) by a single stick, if the O-atom carries a negative charge (–O⁻).

In $(CH_3)_3P{=}O$, with a central phosphorus atom and an overall neutral charge, the terminal O atom must be connected by a double stick to represent a localised double bond (Figure 6.3). Only two of·the four electrons in a double bond actually determine the direction of the double bond in space, as the direction of a double bond is determined by the direction of the underlying single bond. When dealing with poly-atomic oxyanions, such as $SO_4^{2-}$ or $PO_4^{3-}$ (Figure 6.4, examples 8 and 9, respectively), the overall negative charges on the oxyanion, $2-$ in $SO_4^{2-}$ and $3-$ in $PO_4^{3-}$, are localised singly on the available terminal oxygen atoms, and any remaining terminal oxygen atoms are assumed to involve a double bond and are drawn in the ball-and-stick model as a double stick representing a localised double bond. Thus in the $SO_4^{2-}$ oxyanion the double negative charge gives two negative terminal $O^-$ atoms, leaving two further terminal O atoms, each of which must be involved as a double bond to sulfur. In the $PO_4^{3-}$ oxyanion three of the four terminal atoms carry the three negative charges, leaving only one terminal oxygen atom to be involved as a single double bond. Each terminal $O^-$ atom then represents only a single bond to the central atom and contributes just one electron to the valence shell electron count as described in step (c) (ii) of the Working Method (Table 5.2). Each double bond will only contribute one electron for its single bond (or stick), which actually determines the direction of the double bond, but must involve the *subtraction* of an electron for the double bond (or stick), as the second electron pair of the double bond only follows the direction of the underlying single bond, but does not actually determine the direction of the double bond.

### Table 5.2 VSEPR Working Method

*Steps in Using the VSEPR Theory to Predict the Structure of Simple Molecules, Cations and Anions*

(a) Draw a ball-and-stick model of the molecule, anion or cation, with each covalent bond represented as a stick (—), each localised double bond as a double stick (=), and the negative charges on oxyanions localised on the appropriate terminal oxygens.

(b) Circle the central atom about which the stereochemistry is being determined: this defines the valence shell being considered.

(c) Determine the total electron count in the valence shell of the central atom as follows:
   (i) Write down the number of electrons in the valence shell of the central atom ( = group oxidation number, *i.e.* $m + n$ of the $s^m p^n$ valence shell configuration);

(ii) Add *one* electron for each covalent bond (stick or electron pair bond);

(iii) Add *one* electron for *each* delocalised negative charge on an anion such as $BF_4^-$ or $PCl_6^-$, but *not* for negative charges localised on the terminal oxygen atoms;

(iv) Subtract *one* electron for *each* positive charge for a cation, such as $NH_4^+$ or $PCl_4^+$;

(v) Subtract *one* electron for *each* double bond to the central atom.

(d) Divide the total electron count from (c) by two to determine the number of electron pairs contributing to the valence shell of the central atom. Hence, using Table 6.1a or Figure 6.1, determine the spatial arrangement of these electron pairs, *e.g.* two – linear; three – trigonal planar; four – tetrahedral; five – trigonal pyramidal; six – octahedral.

(e) From the number of sticks in (a) and the number of electron pairs in (d), determine the number of bonding pairs and lone pairs.

(f) Draw a *three-dimensional model* of the molecule, anion or cation, including localised double bonds, localised or delocalised charges and significant angles.

## DEVIATIONS FROM REGULAR SHAPES

The regular shapes shown in Figure 6.1 are only realised in practice if the electron pairs are engaged in bonding to identical atoms or groups; *e.g.* all the bonds in methane, $CH_4$, are tetrahedrally arranged, whereas in monochloromethane, $CH_3Cl$, the H–C–H angles are slightly greater (110.9°) than the regular tetrahedral angle of 109.5°, as the more electronegative Cl atom draws the bonding pair of electrons away from the central C atom. Deviations from a regular stereochemistry also occur if lone pairs are present; thus for the tetrahedral stereochemistry associated with four electron pairs it is found by experiment that the ammonia molecule (one lone pair) has a bond angle of 107.4° (Figure 6.2) and the water molecule (two lone pairs) has a bond angle of 104.5°. To account for this latter type of deviation, Gillespie and Nyholm (1957) extended the simple theory, and suggested that repulsion between electron pairs decrease in the order: *lone pair/lone pair > lone pair/bond pair > bond pair/bond pair*. The reason for this is because lone pairs are attracted closer to the nucleus than bonding pairs, as they are subjected to the attraction of a single nucleus, whereas a bonding pair is subjected to the opposing attractions of two nuclei. Consequently, the lone pairs repel each other more strongly causing the two bonding pairs to be forced more closely together. In this way it is possible to explain the decrease in bond angles in passing from methane (no lone pairs) to ammonia (one lone pair) to water (two lone pairs) ($109.5° \rightarrow 107.4° \rightarrow 104.5°$).

Figure 6.4 sets out some examples of the application of the Working Method to the determination of the shape of some molecules, anions and cations using VSEPR theory.

## THE ADVANTAGES OF VSEPR THEORY

- It is unnecessary to remember the shape of every small molecule, anion and cation.
- It is simple to apply using the Working Method.
- It *only* requires a knowledge of the electron configuration of the elements present.
- While only strictly applying to regular stereochemistries, it is readily modified to take account of the presence of lone pairs (l.p.) as distinct from bonding pairs (b.p.) of electrons.
- It emphasises the importance of lone pairs of electrons in determining the distortions from the regular stereochemistries.
- The basis of VSEPR theory does not have to be unlearned later.

## THE DISADVANTAGES OF VSEPR THEORY

- The VSEPR theory cannot be applied to the transition metal complexes where incomplete *d*-shells are involved.
- The more detailed geometry of Main Group elements, especially the heavy elements, involves angular trends, which are difficult to rationalise. For a more advanced discussion of this topic, see R. J. Gillespie, *Chem. Soc. Rev.*, 1992, **21**, 59.

## THE SHAPES OF DINUCLEAR MOLECULES

So far this chapter has shown how the Working Method can be used to determine the shape of mononuclear Main Group molecules, cations and anions, by the application of VSEPR theory to the valence shell $s^m p^n$ electron configuration of the central element present. In practice, the Working Method can equally be applied to determine the stereochemistry of the Main Group elements in polynuclear molecules, anions and cations, with the proviso that the method has to be applied *separately* to each stereochemical centre available in the polynuclear species. Figure 6.5 lists six dinuclear types of organic molecules as stick structures, with the two stereochemical centres circled, to indicate that the Working

| Ball and stick model. | Calculation | Final structure. |
|---|---|---|

**1. Methane**

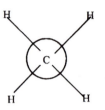

C $s^2p^2$ = 4

4H  1 x 4 = 4

Total = 8 / 2 = 4 el. pr.

*i.e.* tetrahedral;

four bonding pairs

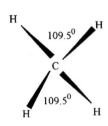

**2. Ammonia.**

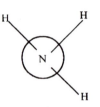

N $s^2p^3$ = 5

3H  1 x 3 = 3

Total = 8 / 2 = 4 el. pr.

*i.e.* tetrahedral; three

bonding + one lone, pairs

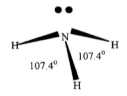

**3. $NH_4^+$ cation.**

N $s^2p^3$ = 5

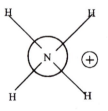

4H  1 x 4 = 4

1 +ve charge = -1

Total = 8 / 2 = 4 el. pr.

*i.e.* tetrahedral;

four  bonding pairs.

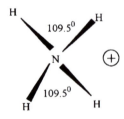

**4  $BF_4^-$ anion.**

B $s^2p^1$ = 3

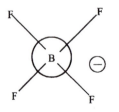

4F  1 x 4 = 4

1 -ve charge = +1

Total = 8 / 2 = 4 el. pr.

*i.e.* tetrahedral;

four bonding pairs.

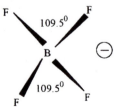

**Figure 6.4** *Examples of applying the Working Method to VSEPR theory*

| Ball and stick model. | Calculation. | Final structure. |
| --- | --- | --- |

5. $(CH_3)_3P{=}O$.

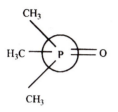

$$P \quad s^2p^3 \qquad = 5$$

$$3C \quad 1 \times 3 \qquad = 3$$

$$1O \quad 1 \times 1 \qquad = 1$$

$$1 \text{ double bond} \quad = -1$$

$$\text{Total} = 8/2 \qquad = 4 \text{ el. pr.}$$

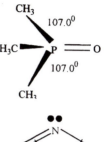

$$1 \text{ double bond} \quad = -1$$

$$\text{Total} = 8/2 = 3 \text{ el. pr.}$$

*i.e.* trigonal planar;

two bonding pairs plus

one double bond and one lone pair.

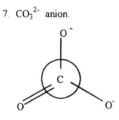

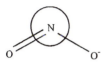

7. $CO_3^{2-}$ anion.

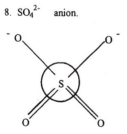

$$C \quad s^2p^2 \qquad = 4$$

$$3O \quad 1 \times 3 \qquad = 3$$

$$1 \text{ double bond} \quad = -1$$

$$\text{Total} = 6/2 = 3 \text{ el. pr.}$$

*i.e.* trigonal planar,

three bonding pairs and one double bond.

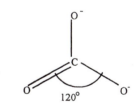

8. $SO_4^{2-}$ anion.

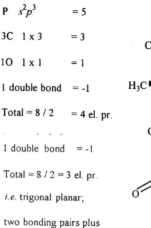

$$S \quad s^2p^4 \qquad = 6$$

$$4O \quad 1 \times 4 \qquad = 4$$

$$2 \text{ double bonds} \quad = -2$$

$$\text{Total} = 8/2 \qquad = 4 \text{ el. pr.}$$

*i.e.* tetrahedral;

four bonding pairs

plus two double bonds.

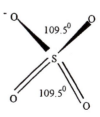

Ball and stick model.                    Calculation.                    Final structure.

9. $PO_4^{3-}$ anion.

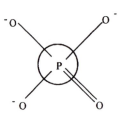

P   $s^2p^3$        = 5

4 O ·1 x 4       = 4

1 double bond = -1

Total = 8 / 2     = 4 el. pr.

*i.e.* tetrahedral;

four bonding pairs,

plus one double bond.

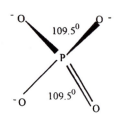

10. $NO_3^-$ anion.

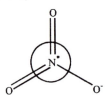

N   $s^2p^3$        = 5

3O  1 x 3        = 3

2 double bonds = -2

Total = 6 / 2 = 3 el. pr.

*i.e.* trigonal planar,

three  bonding pairs

plus two double bonds

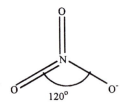

**Note:** There is a problem associated with the representation of the  central nitrogen

atom of the nitrate anion as five coordinate.  This can be avoided as follows:

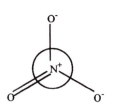

N   $s^2p^3$       = 5            Total = 6 / 2 = 3 el. pr

3O  1 x 3      = 3            *i.e.* trigonal  planar

1 double bond    = -1            three bonding pairs,

1 +ve charge     = -1            one double bond,

                                 one +ve charge.

**Figure 6.4** (*continued*)

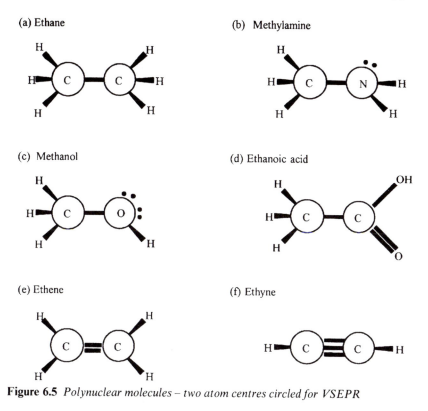

**Figure 6.5** *Polynuclear molecules – two atom centres circled for VSEPR*

Method must be applied separately to these two centres. Thus in the case of ethane both centres are C atoms which have identical tetrahedral stereochemistries, determined in precisely the same way as for the methane molecule in Figure 6.4, example 1, and for this reason the calculation need not be described separately. In the same way, the stereochemistry of any C atom centre in any complex organic molecule may be predicted. However, it must still be remembered that VSEPR theory cannot be used to determine the stereochemistry of transition metal complexes, owing to the presence of an incomplete $d$-subshell.

## HYBRIDISATION OF ATOMIC ORBITALS

VSEPR theory has been used to determine the shape of the methane molecule, Figure 6.4, example 1, as a tetrahedral stereochemistry about the carbon atom. While this simple theory predicts the correct shape, it does this on the basis of a simple numerical count, without considering

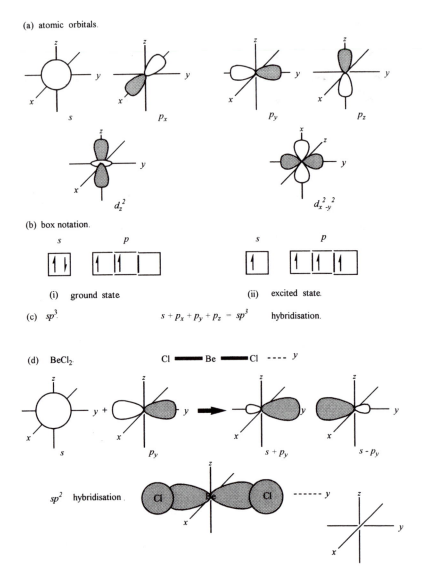

**Figure 6.6** *Hybridisation schemes*

the shape of the orbitals on the central carbon atom, $s^2p^2$, configuration. This involves two electrons in the spherically symmetrical s-orbital, which is then filled, and two electrons in the triply degenerate direction orbitals, $p_x, p_y$ and $p_z$ (Figure 6.6a). Using a box notation to represent the orbitals, the $s^2p^2$ configuration (Figure 6.6b) suggests that only two of these four electrons are unpaired, and available for bond formation. This limitation is removed if the carbon atom is excited to an $s^1 p^3$ configuration (Figure 6.6c), which then has four unpaired electrons available, one $s$- and three $p$-orbitals, for the formation of four C–H bonds. While this is more satisfactory, this excited state configuration still does not predict the formation of four equivalent bonds as suggested by VSEPR theory, and as found in practice. To obtain four equivalent bonds the one $s$-orbital and three $p$-orbitals have to be mixed together to form a hybrid orbital. By taking the appropriate linear combination of atomic orbitals, four equivalent $sp^3$ hybrid orbitals are formed, which, separately, point towards the four corners of a tetrahedron.

$$s + p_x + p_y + p_z \approx sp^3 \text{ hybrid orbitals}$$

The directional properties of the hybrid orbitals are better appreciated with the $BeCl_2$ molecule, which is linear in shape, Cl–Be–Cl 180°, and in which Be has an $s^2$ configuration which can be promoted to an $s^1 p^1$ configuration.

$$\text{Cl–Be–Cl} \rightarrow y\text{-axis}$$

Two linear combinations of the $s$ and $p_y$ orbitals are possible on the central beryllium atom:

$$s + p_y \quad and \quad s - p_y$$

Each hybrid orbital has a better extension along the $\pm y$-axis respectively, and can effect better overlap with the separate orbitals on the Cl(1) and Cl(2) terminal atoms (Figure 6.6d), to form two equivalent covalent bonds, separately, along the $\pm$ y-axis. In the linear combination of atomic orbitals, it is important to specify the particular $p$-orbital involved, in this case the $p_y$-orbital, in an $s + p_y = sp$ hybrid orbital, which then points in the correct direction, $\pm$ y-axes, for effective overlap (Figures 6.6d and 6.7a).

In the same way for the trigonal planar geometry, three planar bonds at 120°, a linear combination of $s + p_x + p_y$ orbitals leads to the formation of *three* equivalent $sp^2$-type orbitals orientated in the $xy$-plane at 120° (Figure 6.7b). For a tetrahedral geometry the linear combination of $s + p_x + p_y + p_z$ is required, generating *four* equivalent $sp^3$-type orbitals (Figure 6.7c). For the five coordinate trigonal bipyramidal geometry the linear combination must involve a $d$-orbital, namely

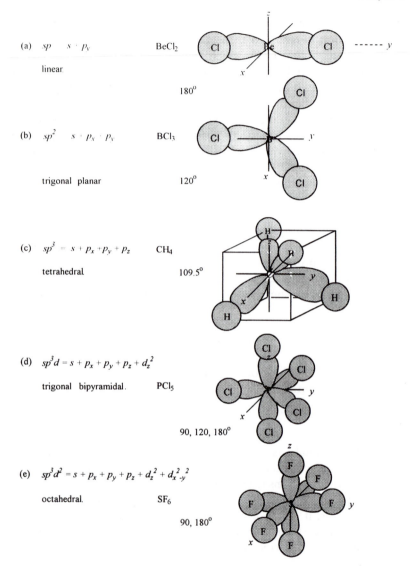

**Figure 6.7** *Basic types of hybridisation*

$s + p_x + p_y + p_z + d_{z^2}$ to generate *five* equivalent $sp^3d$ hybrid orbitals, pointing towards the five corners of a trigonal pyramid, (Figure 6.7d). In this linear combination the $d$-orbital involved is specifically the $d_{z^2}$ orbital, to maximise the bonding along the $z$-axis direction. For the six-coordinate octahedral geometry the linear combination of atomic orbitals must involve two $d$-orbitals, specifically the $d_{z^2}$ and $d_{x^2-y^2}$ orbitals, namely: $s + p_x + p_y + p_z + d_{z^2} + d_{x^2-y^2}$, to yield *six equivalent* $sp^3d^2$-type hybrid orbitals (Figure 6.7e), pointing separately towards the six corners of an octahedron. Figure 6.7 then displays the *five* types of hybridisation, namely, $sp$, $sp^2$, $sp^3$, $sp^3d$ and $sp^3d^2$, that are required to explain the observation of the appropriate number of equivalent bonds in mononuclear molecules.

In all five of these hybrid orbital schemes, the use of hybridisation is only to give an improved directional overlap of orbitals to form two electron pair covalent bonds. *Hybridisation does not determine the basic stereochemistry*. This must still be determined by VSEPR theory and only then can hybridisation schemes be invoked to describe, more effectively, the covalent bonding present. These hybridisation schemes may equally be applied to cations and anions. The $NH_4^+$ cation and $BF_4^-$ anion have already been shown to involve a tetrahedral stereochemistry (Figure 6.4, examples 3 and 4); consequently the bonding in both ions may be described as involving $sp^3$ hybridisation.

Such hybridisation descriptions of the bonding in covalent molecules may also be extended to molecules containing lone pairs of electrons (Figure 6.2), *e.g.* the ammonia molecule $NH_3$. The $NH_3$ molecule has four pairs of electrons in its valence shell, thus determining from VSEPR theory (Figure 6.4, example 2) that its basic geometry is tetrahedral, notwithstanding that three of these electron pairs are involved in N–H covalent bonds and one is involved as a lone pair. This basic tetrahedral geometry then requires four equivalent $sp^3$ type hybrid orbitals, each capable of holding two electrons, as one electron pair, and it does not matter whether the electron pairs in the four $sp^3$ type hybrid orbitals are involved as bonding or lone pairs of electrons. Thus four basic $sp^3$ type hybrid orbitals may be used to describe the bonding in $CH_4$, $NH_3$ or $OH_2$ (Figure 6.8a–c), as they all involve four electron pairs to yield a basic tetrahedral geometry (from VSEPR, Figure 6.4), notwithstanding that they have four, three and two bonding pairs, and zero, one and two lone pairs of electrons, respectively. Equally, as the shapes of the oxy-anions $CO_3^{2-}$ and $SO_4^{2-}$ have been shown in Figure 6.4, examples 7 and 8, to involve a trigonal planar and a tetrahedral stereochemistry, respectively, that is irrespective of the presence of double bonds, their $\sigma$-bonding skeletons may be described by $sp^2$ and the $sp^3$ hybridisations, respectively, that are appropriate to their basic stereochemistry.

The basic hybridisation description of Figure 6.7 can also be used to describe the bonding in transition metal complexes but, as the present text is limited to basic inorganic chemistry, only simple complexes will be considered with a regular octahedral or tetrahedral stereochemistry.

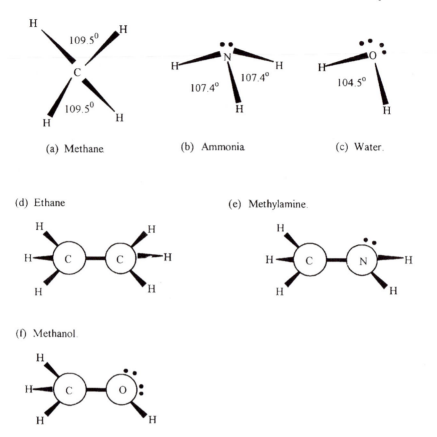

(a) Methane          (b) Ammonia          (c) Water.

(d) Ethane                              (e) Methylamine.

(f) Methanol.

**Figure 6.8** *Tetrahedral centres in some mononuclear and dinuclear molecules*

Thus the $sp^3d^2$ hybridisation description of the bonding in the octahedral stereochemistry of a Main Group compound, *e.g.* $SF_6$, can equally be applied to the bonding in the octahedral stereochemistry of the $[Fe(OH_2)_6]^{2+}$ and $[Mn(OH_2)_6]^{2+}$ cations. Equally, the $sp^3$ hybridisation description of the bonding of a tetrahedral characteristic main group compound, *e.g.* $CH_4$, also describes the bonding in the tetrahedral $MnO_4^-$ and $FeCl_4^-$ anions.

## HYBRIDISATION IN POLYNUCLEAR MOLECULES

The $sp^3$ hybridisation description of the bonding in saturated mononuclear carbon compounds, such as methane, $CH_4$ (Figure 6.8a), may also be extended to saturated polynuclear carbon compounds, such as

(a) Ethene

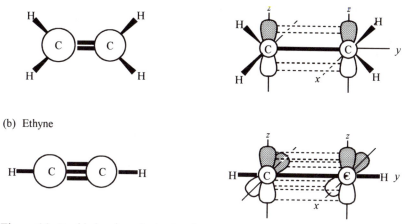

(b) Ethyne

**Figure 6.9** *Double bonds and tripe bonds*

ethane, (Figure 6.8d). In such cases, just as the shape has to be determined at the two separate C atom centres, the hybridisation has also to be determined at each C atom centre and is appropriate to the stereochemistry present. In the case of ethane, both C atoms have the same tetrahedral stereochemistry and consequently the bonding at both carbon atoms can be described by $sp^3$ hybridisation. Equally, the bondings at the C and N atoms of methylamine (Figure 6.8e) and at the C and O atoms of methanol (Figure 6.8f) are all correctly described by $sp^3$ hybridisation, as all four atoms involve a tetrahedral stereochemistry, notwithstanding the presence of one and two lone pairs of electrons on the N and O atoms, respectively.

In the case of unsaturated dinuclear carbon molecules hybridisations other than $sp^3$ are involved at both carbon centres. In ethene (Figure 6.9a), each carbon centre is $sp^2$ hybridised and the three planar $sp^2$ hybrid orbitals bond with two hydrogen atoms and the second carbon atom, leaving a single unhybridised $p_z$-orbital perpendicular to the planes of the $CH_2$ fragments. The three $sp^2$ hybrid orbitals are involved in inplane $\sigma$-bonding, while the pure $p_z$ orbital is involved in $\pi$-bonding perpendicular to the planar $CH_2$ fragment (Figure 6.9a). This generates the double bond description of the ethene carbon–carbon bond, comprising one $\sigma$-bond plus one $\pi$-bond, bond order 2.0 and normally written $H_2C{=}CH_2$. In ethyne (Figure 6.9b), each of the two carbon centres is $sp$ hybridised, and the two linear $sp$ hybrid orbitals form $\sigma$-bonds to one hydrogen and one carbon atom along the $y$-axis, HCCH. This leaves two $p$-orbitals, $p_y$ and $p_z$, perpendicular to the

HCCH direction (Figure 6.9d), available for sideways overlap to form two π-bonds, one in the *xy*-plane and one in the *yz*-plane. This generates the triple bond description of the ethyne carbon to carbon bond, comprising one σ-bond and two π-bonds, bond order 3.0 and normally written as HC≡CH.

## SUMMARY

Given the basic shapes of mononuclear molecules, cations and anions as determined by VSEPR theory, the bonding involved can then be described by the various types of hybrid orbitals, including double and triple bonding. In polynuclear molecules, VSEPR theory can be used to determine the stereochemistry at the *separate* atom centres present. Consequently the bonding at these separate atom centres can still be described by the appropriate types of hybrid orbitals, including multiple bonding.

*Chapter 7*

# A *Features of Interest* Approach to Systematic Inorganic Chemistry

## AIMS AND OBJECTIVES

Previous chapters have shown how the electron configuration of an element determines the Group oxidation number, the variable valence and the physical and chemical properties of the element in its simple compounds. The electron configuration of the central element in a molecule can also be used to describe the Lewis structure, the shape and the bond hybridisation. This chapter extends this approach to show how *80% of the chemistry of simple molecules, anions and cations can be predicted* if the above information derived from the electron configuration of the central element of a molecule is summarised in a *Features of Interest* type Spider Diagram and augmented by information on the preparation, analysis and reactions of the molecule. The *Features of Interest* approach is set out in a Working Method, illustrated by numerous examples, which serves to show students that they know some chemistry and suggests how they write an essay on the structured information of the *Features of Interest* Spider Diagram.

## INTRODUCTION

The position of an element in the Periodic Table is determined by the underlying electron configuration of the element. In particular, the principal quantum number, $n$, largely determines the atomic number, $Z$, and the valence shell configuration of an element determines the vertical Group numbers I–VIII. The Group valence shell configuration determines the Group oxidation number of the element, and also leads to the variable valence of both the Main Group elements and, to a lesser

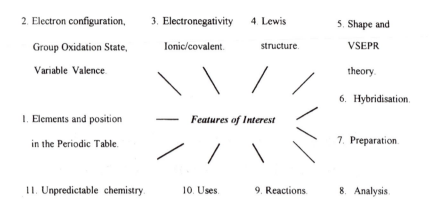

**Figure 7.1**  *A summary of the topics appropriate to a description of the chemistry of simple molecules, a Features of Interest approach*

extent, the Transition Metal elements. The Group oxidation number, combined with the Pauling Electronegativities, determines the formation of ionic or covalent compounds. For the latter, the total number of valence shell electrons, divided by two, decides the number of two-electron pair bonds, which then determines the basic Lewis structure and the shape of the molecule about the central atom by VSEPR theory. Given the shape of a molecule this defines the form of the hybridisation necessary to describe the bonding present. This information is summarised in the top half of the *Features of Interest* Spider Diagram of Figure 7.1.

The present form of the Periodic Table is so clearly based upon the electronic structure of the atom that it is easily forgotten that the original Mendeleev form of the Periodic Table was determined by the different chemical behaviour of the elements. Consequently, given the electron configuration as the basis, a reverse logic may be applied, *i.e.* to use the Periodic Table to predict, in a systematic way, 80% of the chemistry of the elements and their compounds. If, as in the bottom half of Figure 7.1, the additional features of preparation, analysis, reactions and uses are added, such a Spider Diagram goes 80% of the way in predicting the chemistry of a range of simple Main Group and Transition Metal compounds, based upon both the electronic and chemical information inherent in the Periodic Table. A suggested range of simple compounds appropriate to a Basic Introductory Chemistry Course is shown in Table 7.1.

The predictable chemistry of the compounds of Table 7.1 is made up of three parts:

- the method of preparation of the compounds from the parent elements.
- the chemical analysis of the compound, once formed, by acid/base, precipitation or redox reactions.
- the reactions of the compound, on heating and with water.

**Table 7.1** *Some simple inorganic chemistry compounds*

| I | II | III | IV | V | VI | VII |
|---|---|---|---|---|---|---|
| 1 | 2 | 13 | 14 | 15 | 16 | 17 |
| $s^1$ | $s^2$ | $s^2p^1$ | $s^2p^2$ | $s^2p^3$ | $s^2p^4$ | $s^2p^5$ |

*1. Hydrides*

| | | | | | | |
|---|---|---|---|---|---|---|
| LiH | $BeH_2$ | | $CH_4$ | $NH_3$ | $OH_2$ | HF |
| NaH | $MgH_2$ | | $SiH_4$ | $PH_3$ | $SH_2$ | HCl |
| KH | $CaH_2$ | | | | | HBr |
| | | | | | | HI |

*2. Oxides*

| | | | | | | |
|---|---|---|---|---|---|---|
| $Li_2O$ | | | $CO_2$ | $N_2O_5$ | $H_2O$ | |
| | | | CO | $N_2O_3$ | | |
| | | | | $N_2O$ | | |
| $Na_2O$ | MgO | $Al_2O_3$ | $SiO_2$ | $P_2O_5$ | $SO_3$ | $Cl_2O_7$ |
| | | | SiO | $P_2O_3$ | $SO_2$ | |

Transition metal compounds: CuO, $Fe_2O_3$, $MnO_2$

*3. Chlorides*

| | | | | | | |
|---|---|---|---|---|---|---|
| LiCl | $BeCl_2$ | $BCl_3$ | $CCl_4$ | | | |
| NaCl | $MgCl_2$ | $AlCl_3$ | $SiCl_4$ | $PCl_5$ | | |

Transition metal compounds: CuCl, $CuCl_2$, $FeCl_2$, $FeCl_3$

## THE PREPARATION OF SIMPLE COMPOUNDS FROM THE ELEMENTS

Table 7.1 lists approximately 50 simple hydrides, oxides and chlorides of primarily the Main Group elements, which together suggest the 'need to know' 50 separate methods of preparation. Fortunately, this list reduces to heating the corresponding element, solid, liquid or gas, in an atmosphere of hydrogen, oxygen, fluorine or chlorine, respectively, with the stoichiometry of the product determined by the $s^mp^n$ configuration of the element, as set out in Figures 7.2–7.4. In each diagram the oxidation number of the compound formed is plotted against the valence shell configuration of the central element, as this determines the Group oxidation number of the element and hence the stoichiometry of the compound formed. Thus silicon, with a valence shell configuration of $s^2p^2$, yields a Group oxidation number of IV and stoichiometries of $SiH_4$, $SiO_2$ and $SiCl_4$, for the hydride, oxide and chloride, respectively,

(a) First Short Period.

| G.O.N. | I $s^1$ | II $s^2$ | III $s^2p^1$ | IV $s^2p^2$ | V $s^2p^3$ | VI $s^2p^4$ | VII $s^2p^5$ |
|---|---|---|---|---|---|---|---|
| V | | | | | $N_2O_5$ | | |
| IV | | | | $CO_2$ | | | |
| III | | | $B_2O_3$ | | $N_2O_3$ | | |
| II | | BeO | | CO | | | |
| I | $Li_2O$ | | | | | | |
| 0 | Li | Be | B | C | $N_2$ | $O_2$ | $F_2$ |
| -I | | | | | | | FH |
| -II | | | | | | $OH_2$ | |
| -III | | | | | $NH_3$ | | |
| -IV | | | | $CH_4$ | | | |

Oxidation ↑ $O_2$ ; Reduction ↓ $H_2$

(b) Second Short Period.

| G.O.N. | I $s^1$ | II $s^2$ | III $s^2p^1$ | IV $s^2p^2$ | V $s^2p^3$ | VI $s^2p^4$ | VII $s^2p^5$ |
|---|---|---|---|---|---|---|---|
| VII | | | | | | | $Cl_2O_7$ |
| VI | | | | | | | $SO_3$ |
| V | | | | | $P_2O_5$ | | |
| IV | | | | $SiO_2$ | | $SO_2$ | |
| III | | | $Al_2O_3$ | | $P_2O_3$ | | |
| II | | MgO | | | | | |
| I | $Na_2O$ | | | | | | |
| 0 | Na | Mg | Al | Si | $P_4$ | $S_8$ | $Cl_2$ |
| -I | | | | | | | ClH |
| -II | | | | | | $SH_2$ | |
| -III | | | | | $PH_3$ | | |
| -IV | | | | $SiH_4$ | | | |

Oxidation ↑ $O_2$ ; Reduction ↓ $H_2$

**Figure 7.2** *Reactions of the elements: oxidation $(O_2)$ and reduction $(H_2)$*

(a) First Short Period.

| G.O.N. | I $s^1$ | II $s^2$ | III $s^2p^1$ | IV $s^2p^2$ | V $s^2p^3$ | VI $s^2p^4$ | VII $s^2p^5$ |
|---|---|---|---|---|---|---|---|
| VI | | | | | | − | |
| V | | | | | − | | |
| IV | | | | $CF_4$ | | | |
| III | | | $BF_3$ | | | | |
| II | | $BeF_2$ | | | | | |
| 1 | LiF | | | | | | |
| 0 | Li | Be | B | C | $N_2$ | $O_2$ | $F_2$ |

(↑ Oxidation $F_2$)

(b) Second Short Period.

| G.O.N. | I $s^1$ | II $s^2$ | III $s^2p^1$ | IV $s^2p^2$ | V $s^2p^3$ | VI $s^2p^4$ | VII $s^2p^5$ |
|---|---|---|---|---|---|---|---|
| VII | | | | | | | − |
| VI | | | | | | $SF_6$ | |
| V | | | | | $PF_5$ | | |
| IV | | | | $SiF_4$ | | | |
| III | | | $AlF_3$ | | | | |
| II | | $MgF_2$ | | | | | |
| 1 | NaF | | | | | | |
| 0 | Na | Mg | Al | Si | $P_4$ | $S_8$ | $Cl_2$ |

(↑ Oxidation $F_2$)

**Figure 7.3** *Reactions of the elements: oxidation ($F_2$)*

notwithstanding, the presence of formal oxidation number of − IV for the hydride and IV for the oxide and chloride, respectively. The formation of a formal negative − IV oxidation number from the elemental silicon is then a process of reduction, the addition of four electrons to give an $s^2p^6$ closed shell configuration, while the formation of a formal positive IV oxidation state from the elemental silicon is a process of oxidation, the removal of four electrons to give an $s^0p^0$ valence shell configuration, with the difference, − IV to IV, equal to the eight electrons of an inert gas valence shell configuration (see Figure 4.2).

In the oxidation process, oxygen and chlorine, the positive product is in the Group oxidation number, for the Main Group elements whereas for the reduction process, the negative product is 8 − GON, where the

(a) First Short Period.

| G.O.S. | I $s^1$ | II $s^2$ | III $s^2p^1$ | IV $s^2p^2$ | V $s^2p^3$ | VI $s^2p^4$ | VII $s^2p^5$ |
|---|---|---|---|---|---|---|---|
| VI |  |  |  |  |  | - |  |
| V |  |  |  |  | - |  |  |
| IV |  |  |  | $CCl_4$ |  |  |  |
| III |  |  | $BCl_3$ |  | $NCl_3$ |  |  |
| II |  | $BeCl_2$ |  |  |  |  |  |
| I | $LiCl$ |  |  |  |  |  |  |
| 0 | Li | Be | B | C | $N_2$ | $O_2$ | $F_2$ |

Oxidation $Cl_2$ (↑)

(b) Second Short Period.

| Oxidation $Cl_2$ | I $s^1$ | II $s^2$ | III $s^2p^1$ | IV $s^2p^2$ | V $s^2p^3$ | VI $s^2p^4$ | VII $s^2p^5$ |
|---|---|---|---|---|---|---|---|
| VII |  |  |  |  |  |  | - |
| VI |  |  |  |  |  | - |  |
| V |  |  |  |  | $PCl_5$ |  |  |
| IV |  |  |  | $SiCl_4$ |  | $SCl_4$ |  |
| III |  |  | $AlCl_3$ |  | $PCl_3$ |  |  |
| II |  | $MgCl_2$ |  |  |  | $SCl_2$ |  |
| I | $NaCl$ |  |  |  |  |  |  |
| 0 | Na | Mg | Al | Si | $P_4$ | $S_8$ | $Cl_2$ |

**Figure 7.4** *Reactions of the elements: oxidation ($Cl_2$)*

positive and negative oxidation numbers are connected by a difference of $s^2p^6$, namely, an outer valence electron shell of eight electrons, *i.e.* $C^{IV}O_2$ and $C^{-IV}H_4$. In some of the Main Group elements, positive oxidation numbers lower than the GON occur, owing to the stability of the $s^2$ *pseudo* inert gas core configuration, as in the $C^{IV}O_2$ ($s^2p^2$) and $C^{II}O$ ($s^2p^0$) configurations. The oxidation process, *via* chlorine, can be extended to the other halogens. With the most electronegative element fluorine, the highest oxidation state of an element can be predicted, *e.g.* the GON of $S^{VI}F_6$. With the less electronegative halogens like iodine, lower oxidation states than the GON can be produced; thus chlorine produces $S^{IV}Cl_4$. With the transition metals, heating with the halogens rarely produces the GON, even with fluorine, and iodine only produces

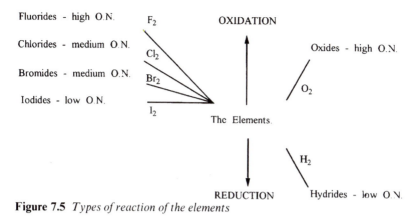

**Figure 7.5** *Types of reaction of the elements*

a lower oxidation number. Thus heating iron with the halogens results in the formation of $Fe^{IV}F_4$ with fluorine, $Fe^{III}Cl_3$ and $Fe^{III}Br_3$, with chlorine and bromine, respectively, but only $Fe^{II}I_2$ with iodine. Likewise, the addition of KCl to an aqueous solution of $[Cu^{II}(OH_2)_6]SO_4$, produces only $K_2[Cu^{II}Cl_4(OH_2)_2]$, while the addition of KI to the same solution reduces copper(II) to copper(I), with the precipitation of $Cu_2I_2$ and the liberation of free iodine, $I^0$ (Figure 7.5).

In the addition of a transition metal to an aqueous solution of an acid, the product is very dependent on the nature of the acid. With iron metal and the oxidising nitric acid, $HNO_3$; iron(III) as $[Fe^{III}(OH_2)_6](NO_3)_3$ is produced, but with the non-oxidising sulfuric acid, $H_2SO_4$, only iron(II) is produced, as $[Fe^{II}(OH_2)_6]SO_4$.

$$Fe + H_2SO_4 + 6H_2O \rightarrow [Fe(OH_2)_6]SO_4 + H_2 \uparrow$$

## THE REACTIONS OF SIMPLE COMPOUNDS

Having produced the compounds of Table 1, their reactions can be explored in the following ways, as shown in Figure 7.6:

- How do they react with water?
- Can they be analysed by the reactions of simple volumetric chemistry, namely, acid/base, precipitation or redox reactions?
- How do the compounds behave on heating?

### Reaction with Water

Most of the oxides react with water to give a base (Groups I and II),

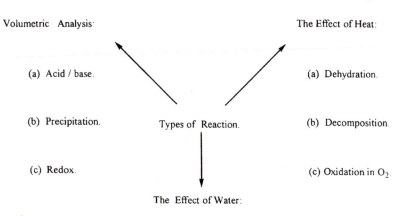

**Figure 7.6** *Types of reaction of simple compounds of the elements*

$$Li_2O + H_2O \rightarrow 2LiOH$$

or an oxyacid (Groups III–VII; see Figure 7.7):

$$CO_2 + H_2O \rightarrow H_2CO_3$$

In general, the halides either simply ionise or hydrolyse to the parent oxyacid:

$$PCl_5 + 4H_2O \rightarrow H_3PO_4 + 5HCl$$

The reaction of a simple compound with water usually yields products that can be analysed by volumetric chemistry, as a second phase reaction. Thus the reaction of $PCl_5$ with water yields HCl and $H_3PO_4$, both of which can be titrated with NaOH solution, and HCl with $AgNO_3$ solution.

## Volumetric Reactions

A valuable source of knowledge on chemical reactions can be found in the quantitative and qualitative reactions of analytical chemistry, which a student meets early in the practical chemistry laboratory. These involve a range of experience of simple, and not so simple, acid/base,

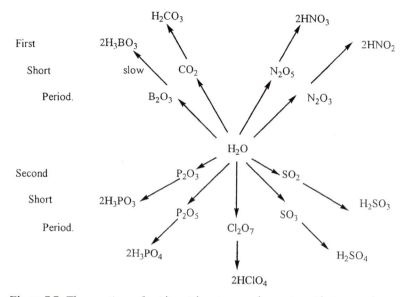

**Figure 7.7** *The reactions of oxides with water, no change in oxidation number*

precipitation and redox reactions, usually carried out in water, which form a useful initial data base of reactions that the student can add to with experience. As all of the methods of simple volumetric chemistry involve reactions in aqueous solution, these analyses form a useful source of chemical reactions of the simple compounds of Table 7.1. Thus:

$$\text{Acid/base: } H_2CO_3 + 2NaOH \rightarrow Na_2CO_3 + 2H_2O$$
$$\text{Precipitation: } NaCl + AgNO_3 \rightarrow AgCl\downarrow + NaNO_3$$

$$\text{Redox: } 5Fe^{II}Cl_2 + KMn^{VII}O_4 \rightarrow 5[Fe^{III}(OH_2)_6]^{2+} + [Mn^{II}(OH_2)_6]^{2+}$$

For this reason each component element of a compound should be examined separately for analysis by one of the three basic types of volumetric reactions, namely acid/base, precipitation or redox analysis, as these represent a useful source of information on chemical reactions, familiar from laboratory practical work.

### The Effect of Heat

Another potential source of information on chemical reactions is the effect of heat on the simple compounds of Table 7.1. At its simplest, if the compounds are hydrated, heating can produce the anhydrous compound, a process that can be used for *thermogravimetric analysis*:

$$BaCl_2 \cdot 2H_2O \rightarrow BaCl_2 + 2H_2O$$

The effect of heating may be to produce a *decomposition*, in which either no change in the oxidation number occurs, *i.e.*

$$Ca^{II}C^{IV}O_3 \rightarrow Ca^{II}O + C^{IV}O_2$$

or in which some change in the oxidation number occurs, *i.e.*

$$2NaN^{V}O_3 \rightarrow 2NaN^{III}O_2 + O_2$$

In this latter reaction, the initial and final oxidation states of the nitrogen atoms differ by the two electrons of the $s^2$ *pseudo* inert gas core configuration.

Another useful general reaction is that the hydrides, when heated in oxygen gas, are oxidised, *i.e.*

$$C^{-IV}H_4 + 2O_2 \rightarrow C^{IV}O_2 + 2H_2O \rightarrow H_2C^{IV}O_3$$

an oxidation process involving the 8e change, $-$ IV to IV.

To be systematic about the 'predictable' chemistry, it is useful to have a Working Method for this *Features of Interest* approach to systematic inorganic chemistry, which serves the purpose of prodding the student into asking a number of relevant questions that will encourage an intelligent guess at the answer, rather than have to memorise a vast amount of factual information.

### *FEATURES OF INTEREST* OF SIMPLE COMPOUNDS –
### WORKING METHOD

Figure 7.1 sets out a *Features of Interest* approach to the properties of the simple compounds of Table 7.1, and there follows a suggested set of questions as a Working Method to this approach.

Q.1. Which elements are present, to which block in the Periodic Table do they belong, and what are their Valence Shell electron configurations?
Q.2. What are their oxidation numbers, and do they exhibit variable valence, as in their V diagrams (Figure 4.2)?
Q.3. From Pauling's relative electronegativities (EN) (Table 4.2) of the

elements, what type of bonding in the given compound is predicted, *i.e.* ionic or covalent? Use a rule of thumb approach, *i e* an EN difference > 0.9 suggests an ionic bond, an EN difference of < 0.9 suggests covalent bonding. In general, a covalent molecule has shape.

Q.4. Predict the Lewis structure of the simple compound for both ionic and covalent structures, comment on a non-octet configuration and estimate the formal charge on each atom, where appropriate.

Q.5. Predict the shape of the covalent molecules by VSEPR theory and draw a ball-and-stick diagram to represent the structure, including regular or distorted angles.

Q.6. Predict the hybridisation involved appropriate to the above structure, and list the linear combination of atomic orbitals involved by specifying the number of individual *s*-, *p*- and *d*-orbitals involved and, if possible, draw a sketch of the resultant hybrid orbital.

Q.7. Use the information of Figures 7.2–7.7 to suggest a method of preparation of the compound directly from the free component elements, and use some imagination.

Q.8. How would you obtain an analysis of the compound: (a) acid/base; (b) precipitation; (c) redox reactions?

Q.9. What sort of reactions does the compound undergo? Use the data of Figure 7.6 to suggest possible types of reaction, and use some imagination.

Q.10. What uses does the compound have: (a) in the laboratory; (b) in industry?

Q.11. Does the compound have any 'unpredictable chemistry'?

## THE APPLICATION OF THE WORKING METHOD TO A SELECTION OF SIMPLE COMPOUNDS

In using the Working Method above every attempt should be made to relate the inorganic chemistry properties to the chemistry of the various sections (1–11) in the eight **Spider Diagrams** of Figures 7.8a–h.

In the light of these examples the following comments should be noted.

1. The application of the Working Method to *all* the compounds is not equally successful; it is most successful with simple covalent compounds.
2. Although logically the preparation, Q.7, could be thought to come first, working through the answers to Q.1–6 (which are purely deductive from the elements present) may be of some help in predicting or reminding the student of a preparation or a reaction.

**Table 7.2** *Abbreviated table of electronegativities ( Pauling's values)*

| H | | | | | | |
|---|---|---|---|---|---|---|
| 2.1 | | | | | | |
| Li | Be | B | C | N | O | F |
| 1.0 | 1.5 | 2.0 | 2.5 | 3.0 | 3.5 | 4.0 |
| Na | Mg | Al | Si | P | S | Cl |
| 0.9 | 1.2 | 1.5 | 1.8 | 2.1 | 2.5 | 3.0 |
| K | Ca | | | | | Br |
| 0.8 | 1.1 | | | | | 2.75 |
| First row transition metals | | | | | | I |
| 1.2–1.75 | | | | | | 2.2 |

3. Probably the most difficult question is whether the bonding is ionic or covalent. In general, the *s*-block elements (as cations with an inert gas configuration) are ionic, *p*-block elements (excluding the most electronegative elements of Groups VI and VII) are covalently bonded. At room temperature ionic compounds are usually solids and covalent compounds are either gases or liquids depending on the polarity of the molecules.

4. In determining the shape by VSEPR theory, draw a 'ball-and-stick model' first, and circle the element whose stereochemistry is to be determined, just to make this clear.

5. The inclusion of a question on analysis may at first sight seem unfair, but most of the possible reactions will already have been dealt with in other sections of the syllabus, namely acid/base, precipitation and redox reactions in the practical course, and need only be applied to the specific compound. Likewise, it is also useful to emphasise the role of inorganic compounds as reagents in organic chemistry, such as $PCl_5$, $FeCl_3$ and $AlCl_3$, at the appropriate stage.

6. In nearly all the examples, there is an unpredictable element (20%) in the deductive chemistry, but this is not so important that the predictable chemistry is invalidated.

The use of the *Features of Interest* Spider Diagram can readily be extended to separate cations, such as the $NH_4^+$ cation and to anions, such as the $BF_4^-$ anion, which when taken together provide a *Features of Interest* approach to a salt, *i.e.* $NH_4BF_4^-$. The approach can also be used to describe the chemistry of oxyacids, such as $H_2CO_3$ (Figure 7.8h) or their anions, $CO_3^{2-}$. The chemistry of simple transition metal compounds can be described, such as $[Fe(OH_2)_6]SO_4$ (Figure 7.8g), for which additional information is available from the *Features of Interest* description of the chemistry of the separate $OH_2$ ligands and the $SO_4^{2-}$ anions.

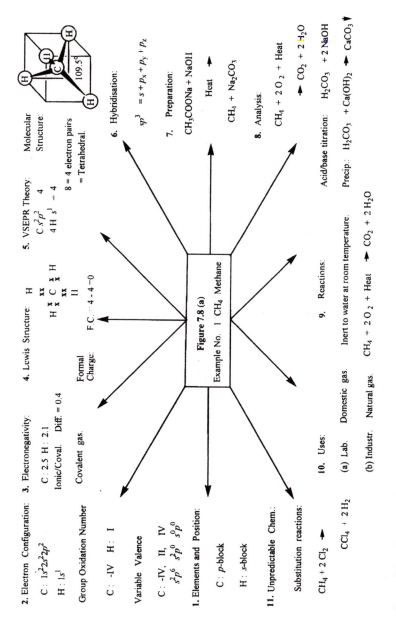

**2. Electron Configuration:**

$C : 1s^2 2s^2 2p^2$

$H : 1s^1$

**3. Electronegativity:**

$C : 2.5$   $H : 2.1$

Ionic/Coval.   Diff. $= 0.4$

Covalent gas.

**4. Lewis Structure:**

H
xx
H $\overset{x}{\underset{x}{C}} \overset{xx}{\underset{xx}{H}}$ H
II

Formal Charge:   F.C. $-4-4-0$

**5. VSEPR Theory:**

$C s^2 p^2$   4

$4 H s^1 - 4$

$8 = 4$ electron pairs
$=$ Tetrahedral.

**Molecular Structure:**

109.5°

Group Oxidation Number

$C : -IV$   $H : I$

Variable Valence

$C : -IV, II, IV$
$s^2 p^6$   $s^2 p^0$   $s^0 p^0$

**1. Elements and Position:**

$C : p$-block

$H : s$-block

**11. Unpredictable Chem.:**

Substitution reactions:

$CH_4 + 2 Cl_2 \rightarrow$

$CCl_4 + 2 H_2$

**6. Hybridisation:**

$sp^3 = s + p_x + p_y + p_z$

**7. Preparation:**

$CH_3COONa + NaOH$

$\xrightarrow{\text{Heat}}$

$CH_4 + Na_2CO_3$

**8. Analysis:**

$CH_4 + 2 O_2 + Heat$

$\rightarrow CO_2 + 2 H_2O$

Acid/base titration:   $H_2CO_3 + 2 NaOH$

Precip.:   $H_2CO_3 + Ca(OH)_2 \rightarrow CaCO_3$ ▼

**10. Uses:**

(a) Lab.   Domestic gas.

(b) Industr.   Natural gas.

**9. Reactions:**

Inert to water at room temperature.

$CH_4 + 2 O_2 + Heat \rightarrow CO_2 + 2 H_2O$

**Figure 7.8 (a)**

Example No. 1 $CH_4$ Methane

**Figure 7.8a** *Features of Interest Spider Diagram for methane*

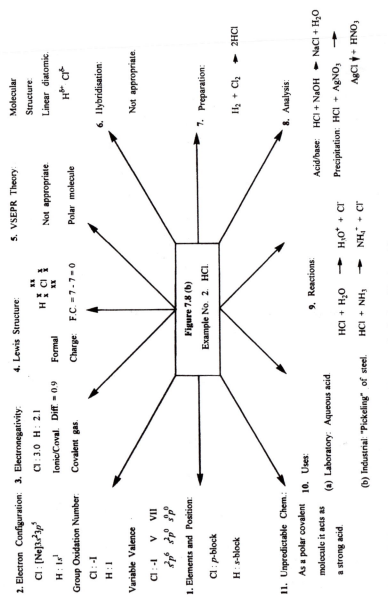

**2. Electron Configuration:**

Cl : [Ne]$3s^2 3p^5$

H : $1s^1$

**Group Oxidation Number**

Cl : -I

H : I

**Variable Valence**

Cl : **-I      V      VII**

$s^2 p^6$   $s^2 p^0$   $s^0 p^0$

**1. Elements and Position:**

Cl : p-block

H : s-block

**11. Unpredictable Chem.:**

As a polar covalent
molecule it acts as
a strong acid.

**3. Electronegativity:**

Cl : 3.0   H : 2.1

Ionic/Coval.   **Diff. = 0.9**

Covalent gas.

**4. Lewis Structure:**

H $\overset{x}{\underset{x}{\times}}$ Cl $\overset{xx}{\underset{xx}{\times}}$

**Formal**        **Charge:**   F.C. = 7 - 7 = 0

**5. VSEPR Theory:**

Not appropriate.

Polar molecule

**6. Hybridisation:**

Not appropriate.

**Molecular Structure:**

Linear diatomic.

H$^{\delta+}$  Cl$^{\delta-}$

**7. Preparation:**

$H_2 + Cl_2 \rightarrow 2HCl$

**8. Analysis:**

Acid/base:   HCl + NaOH $\rightarrow$ NaCl + $H_2O$

Precipitation: HCl + $AgNO_3$ $\uparrow$

AgCl $\downarrow$ + $HNO_3$

**9. Reactions:**

HCl + $H_2O$ $\rightarrow$ $H_3O^+$ + Cl$^-$

HCl + $NH_3$ $\rightarrow$ $NH_4^+$ + Cl$^-$

**10. Uses:**

(a) Laboratory: Aqueous acid.

(b) Industrial: "Pickeling" of steel.

**Figure 7.8 (b)**

Example No. 2.   HCl.

**Figure 7.8b** *Features of Interest Spider Diagram for hydrochloric acid*

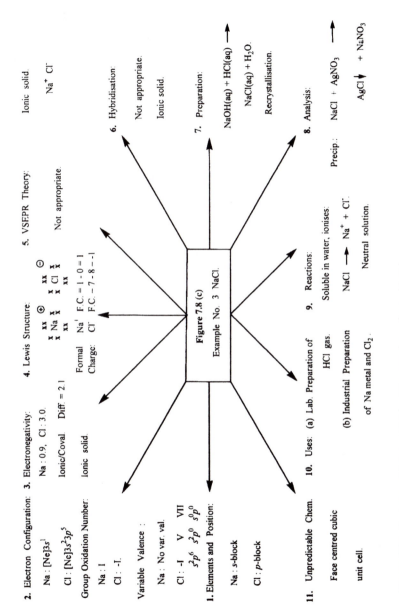

**2. Electron Configuration:**

Na : [Ne]$3s^1$

Cl : [Ne]$3s^2 3p^5$

**Group Oxidation Number:**

Na : I

Cl : -I.

Variable Valence :

Na : No var. val.

Cl : -I    V    VII
$s^2 p^6$  $s^2 p^0$  $s^0 p^0$

**1. Elements and Position:**

Na : s-block

Cl : p-block

**3. Electronegativity:**

Na : 0.9,   Cl : 3.0.

Ionic/Coval.   Diff. = 2.1

Ionic solid.

**4. Lewis Structure:**

xx   ⊕        xx   ⊖
x Na x       x Cl x
xx            xx

Formal   Na⁺  F.C. = 1 - 0 = 1
Charge:  Cl⁻  F.C. - 7 - 8 - -1

**Figure 7.8 (c)**

Example No. 3  NaCl.

**5. VSEPR Theory:**

Not appropriate.

Ionic solid.

Na⁺ Cl⁻

**6. Hybridisation:**

Not appropriate.

Ionic solid.

**7. Preparation:**

NaOH(aq) + HCl(aq) →

NaCl(aq) + H₂O.

Recrystallisation.

**8. Analysis:**

Precip.:   NaCl + AgNO₃ →

AgCl↓   + NaNO₃

**9. Reactions:**

Soluble in water, ionises:

NaCl → Na⁺ + Cl⁻.

Neutral solution.

**10. Uses: (a) Lab. Preparation of**

HCl gas.

**(b) Industrial Preparation**

of Na metal and Cl₂.

**11. Unpredictable Chem.**

Face centred cubic

unit cell.

**Figure 7.8c** *Features of Interest Spider Diagram for sodium chloride*

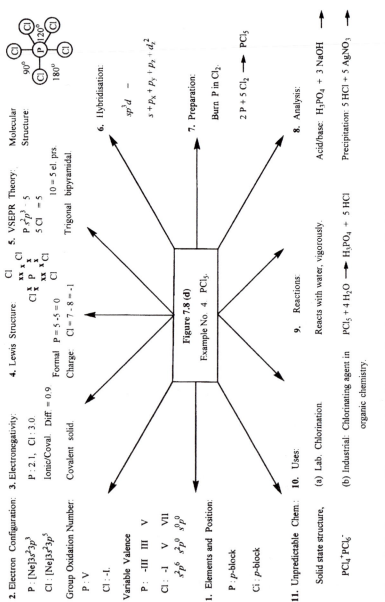

**2. Electron Configuration:**

P : [Ne]$3s^23p^3$

Cl : [Ne]$3s^23p^5$

Group Oxidation Number:

P : V

Cl : -I.

Variable Valence

P :   -III   III   V

Cl :   -I   V   VII

   $s^2p^6$   $s^2p^0$   $s^0p^0$

**1. Elements and Position:**

P : *p*-block

Cl : *p*-block

**11. Unpredictable Chem.:**

Solid state structure,

$PCl_4^+PCl_6^-$

**3. Electronegativity:**

P : 2.1,   Cl : 3.0.

Ionic/Coval.   Diff = 0.9

Covalent solid.

**4. Lewis Structure**

$$\text{Cl}\overset{\times}{\underset{\times}{\text{x}}}\overset{\text{Cl}}{\underset{\text{x}}{\overset{\text{x}}{\text{P}}}}\overset{\text{Cl}}{\underset{\times\times}{\text{x x}}}\text{Cl}$$

Formal   P = 5 - 5 = 0

Charge:   Cl = 7 - 8 = -1

**5. VSEPR Theory:**

P : $s^2p^3$ - 5

5 Cl = 5

10 = 5 el. prs.

Trigonal bipyramidal.

Molecular Structure:

**6. Hybridisation:**

$sp^3d$ –

$s + p_x + p_y + p_z + d_{z^2}$

**7. Preparation:**

Burn P in $Cl_2$.

$2 P + 5 Cl_2 \longrightarrow PCl_5$

**8. Analysis:**

Acid/base: $H_3PO_4 + 3$ NaOH

Precipitation: 5 HCl + 5 $AgNO_3$

**9. Reactions:**

Reacts with water, vigorously.

$PCl_5 + 4 H_2O \longrightarrow H_3PO_4 + 5$ HCl

**10. Uses:**

(a) Lab. Chlorination.

(b) Industrial: Chlorinating agent in organic chemistry.

Figure 7.8 (d)

Example No. 4.   $PCl_5$.

**Figure 7.8d**   *Features of Interest Spider Diagram for phosphorus pentachloride*

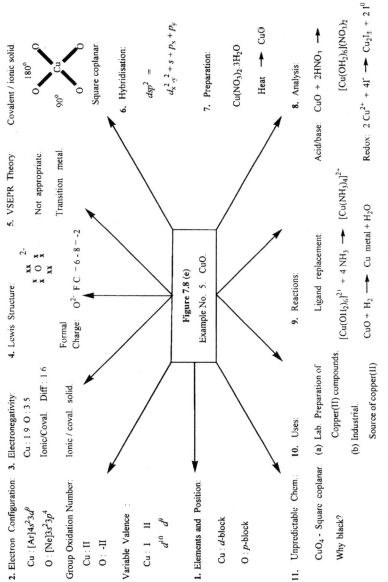

**2. Electron Configuration:**

Cu: [Ar]$4s^2 3d^9$

O: [Ne]$3s^2 3p^4$

Group Oxidation Number:

Cu: II

O: -II

Variable Valence :

Cu: I   II

    $d^{10}$   $d^9$

**3. Electronegativity:**

Cu: 1.9   O: 3.5

Ionic/Coval.   Diff.: 1.6

Ionic / coval. solid.

**4. Lewis Structure:**

Formal Charge:   O$^{2-}$   F.C. - 6 - 8 - -2

**5. VSEPR Theory:**

Not appropriate.

Transition metal.

Covalent / ionic solid.

$180^0$

$90^0$

Square coplanar.

**6. Hybridisation:**

$dsp^2 =$

$d_{x^2-y^2}^2 + s + p_x + p_y$

**7. Preparation:**

$Cu(NO_3)_2.3H_2O$

Heat   $\longrightarrow$   CuO

**8. Analysis:**

Acid/base:   CuO + 2HNO$_3$   $\longrightarrow$

[Cu(OH$_2$)$_6$](NO$_3$)$_2$

Redox:   2 Cu$^{2+}$ + 4I$^-$   $\longrightarrow$   Cu$_2$I$_2$ + 2 I$^0$

**1. Elements and Position:**

Cu: *d*-block

O: *p*-block

**9. Reactions:**

Ligand replacement:

[Cu(OH$_2$)$_6$]I$^{2+}$ + 4 NH$_3$   $\longrightarrow$   [Cu(NH$_3$)$_4$]$^{2+}$

CuO + H$_2$   $\longrightarrow$   Cu metal + H$_2$O

**11. Unpredictable Chem.:**

CuO$_4$ - Square coplanar.

Why black?

**10. Uses:**

(a) Lab. Preparation of Copper(II) compounds.

(b) Industrial.

Source of copper(II)

**Figure 7.8 (e)**

Example No. 5. CuO.

**Figure 7.8e** *Features of Interest Spider Diagram for copper(II) oxide*

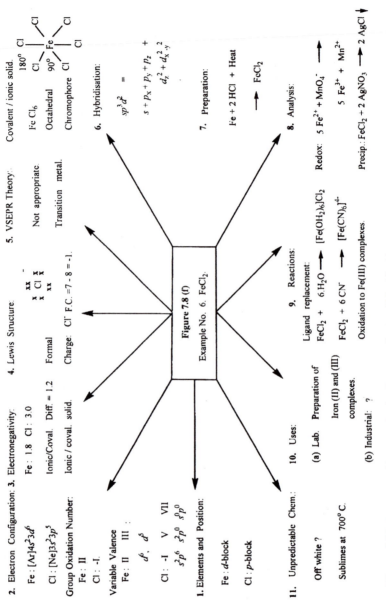

**Figure 7.8f** *Features of Interest Spider Diagram for iron(II) chloride*

2. Electron Configuration:
Fe : [Ar]$4s^2 3d^6$
Cl : [Ne]$3s^2 3p^5$

3. Electronegativity:
Fe : 1.8   Cl : 3.0
Ionic/Coval.  Diff. = 1.2

Group Oxidation Number:
Fe : II
Cl : -I.

Variable Valence
Fe : II   III :
    $d^6$, $d^5$
Cl : -I   V   VII
    $s^2 p^6$  $s^2 p^0$  $s^0 p^0$

1. Elements and Position:
Fe : *d*-block
Cl : *p*-block

11. Unpredictable Chem.:
Off white ?
Sublimes at 700° C.

4. Lewis Structure:

xx
x Cl x
xx

Formal

Charge: Cl⁻  F.C. = 7 - 8 = -1.

5. VSEPR Theory:
Not appropriate
Transition metal.

Covalent / ionic solid:
Fe Cl₆
Octahedral
Chromophore

6. Hybridisation:
$sp^3 d^2$ =
$s + p_x + p_y + p_z + d_z^2 + d_{x^2-y^2}$

7. Preparation:
Fe + 2 HCl + Heat
       →  FeCl₂

8. Analysis:
Redox:  5 Fe²⁺ + MnO₄⁻  →  5 Fe³⁺ + Mn²⁺

Precip.: FeCl₂ + 2 AgNO₃  →  2 AgCl

9. Reactions:
Ligand replacement:
FeCl₂ + 6.H₂O  →  [Fe(OH₂)₆]Cl₂
FeCl₂ + 6 CN  →  [Fe(CN)₆]⁴⁻
Oxidation to Fe(III) complexes.

10. Uses:
(a) Lab.  Preparation of
Iron (II) and (III)
complexes.
(b) Industrial: ?

**Figure 7.8 (f)**
Example No. 6. FeCl₂.

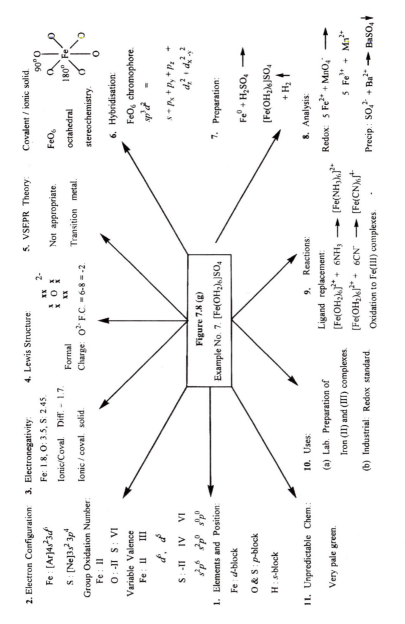

**Figure 7.8g** *Features of Interest Spider Diagram for hydrated iron(II) sulfate hexahydrate*

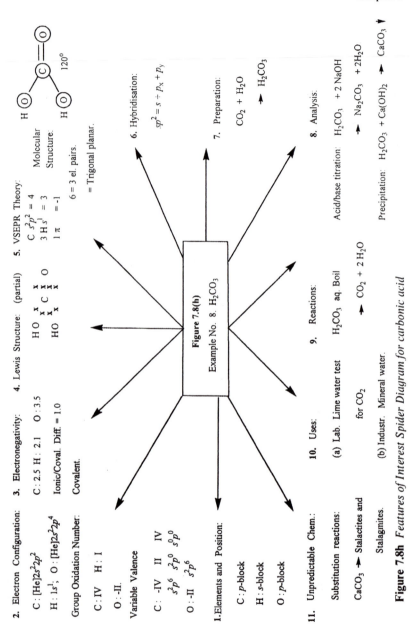

**2. Electron Configuration:**

C : [He]$2s^2 2p^2$   O : [He]$2s^2 2p^4$

H : $1s^1$,   O : [He]$2s^2 2p^4$

**Group Oxidation Number:**

C : IV   H : I

O : -II.

**Variable Valence**

C :  -IV   II   IV
     $s^2 p^6$  $s^2 p^0$  $s^0 p^0$

O : -II   $s^2 p^6$

**1. Elements and Position:**

C : $p$-block

H : $s$-block

O : $p$-block

**3. Electronegativity:**

C : 2.5   H : 2.1   O : 3.5

Ionic/Coval. Diff = 1.0

Covalent.

**4. Lewis Structure:** (partial)

H O  ×× C ××  O
H O  ×× ×× ××

**5. VSEPR Theory:**

C $s^2 p^2 = 4$
3 H $s^1 = 3$
1 π  = -1

6 = 3 el. pairs.

= Trigonal planar.

Molecular Structure:

120°

**6. Hybridisation:**

$sp^2 = s + p_x + p_y$

**7. Preparation:**

$CO_2 + H_2O$

↓

$H_2CO_3$

**8. Analysis:**

Acid/base titration:  $H_2CO_3 + 2\ NaOH$

↓

$Na_2CO_3 + 2\ H_2O$

Precipitation:  $H_2CO_3 + Ca(OH)_2$ → $CaCO_3$ ▼

**9. Reactions:**

$H_2CO_3$ aq. Boil

↓

$CO_2 + 2\ H_2O$

**10. Uses:**

(a) Lab. Lime water test for $CO_2$

(b) Industr. Mineral water.

**11. Unpredictable Chem.:**

Substitution reactions:

$CaCO_3$ → Stalactites and Stalagmites.

**Figure 7.8(h)**

Example No. 8. $H_2CO_3$

**Figure 7.8h** *Features of Interest Spider Diagram for carbonic acid*

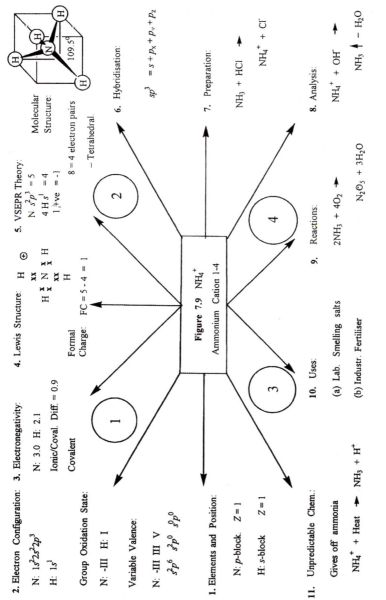

**Figure 7.9** *Features of Interest Spider Diagram for the ammonium cation, subdivided into four paragraphs*

## WRITING AN ESSAY OR REPORT FROM THE SPIDER DIAGRAM

The *Features of Interest* Spider Diagram builds up a substantial amount of simple chemical information, 80% of which is predictable from a knowledge of the electron configuration of the central element. But in addition the information is presented in a structured way that can be used to write a report or essay on the chemistry of the molecule, cation or anion of the *Features of Interest* approach. Thus the Spider Diagram of $NH_4^+$ can be divided into four quadrants (Figure 7.9) to form the four main paragraphs of an essay or report, which with the addition of Introduction and Conclusion paragraphs (Figure 7.10) would complete the formal structure of an essay or report on the chemistry of the $NH_4^+$ cation. Given the outline for the essay, the following section gives an example:

*Introduction. The ammonium cation has the formula $NH_4^+$ and is one of the simplest inorganic polyatomic cations that is readily available in the laboratory as ammonium hydroxide, $NH_4OH$, aqueous solution.*

*Paragraph 1. The ammonium cation is made up of one atom of nitrogen and four atoms of hydrogen. Nitrogen has an atomic number of 7 and is a p-block element in Group V, row 2 with an electron configuration $1s^2 2s^2 2p^3$. Hydrogen has an atomic number of 1 and is an s-block element in Group 1 row 1, with an electron configuration $1s^1$. The $NH_4^+$ cation has an overall single positive charge. The nitrogen has a formal oxidation number of $-III$ and the hydrogen one of I, with electron configurations of $1s^2 2s^2 2p^6$ and $1s^0$, respectively, in these oxidation states. With an electronegativity of 3.00 for nitrogen and 2.1 for hydrogen, the difference of 0.90 suggests a covalent molecule.*

*Paragraph 2. The ammonium cation has a Lewis structure involving eight electrons in the valence shell, arranged as four electron pairs bonding the central nitrogen atom to the four terminal hydrogen atoms and forming four bonding pairs of electrons. The shape of the $NH_4^+$ cation can be determined by VSEPR theory; the central nitrogen atom provides five electrons from its valence shell configuration, $s^2 p^3$ and the four separate hydrogen atoms provide one electron each from their $s^1$ electron configuration. This gives a total of nine electrons. Subtracting one electron for the positive charge leaves a total of eight electrons, and dividing by 2 gives four electron pairs and predicts the stereochemistry of the $NH_4^+$ cation to be a*

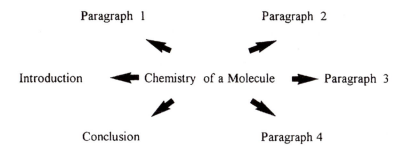

Figure 7.10 *A Spider Diagram for a basic essay structure*

regular tetrahedron with a bond angle of 109.5°. The bonding in the four tetrahedral N–H bonds of the $NH_4^+$ cation is best described as involving $sp^3$ type hybridisation.

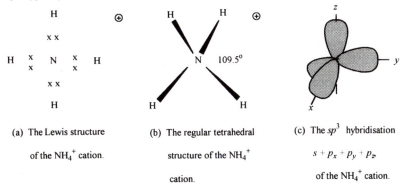

(a) The Lewis structure of the $NH_4^+$ cation.

(b) The regular tetrahedral structure of the $NH_4^+$ cation.

(c) The $sp^3$ hybridisation $s + p_x + p_y + p_z$ of the $NH_4^+$ cation.

**Paragraph 3.** The ammonium cation can be prepared as one of the products of the reaction of ammonia with an acid:

$$NH_3 + HCl \rightarrow NH_4^+ + Cl^-$$

It can be analysed by adding NaOH solution, boiling off the $NH_3$ into standard acid and back titrating with standard alkali:

$$NH_4^+ + OH^- \rightarrow NH_3 + H_2O$$

The $NH_4^+$ cation can be oxidised in air according to the following reaction:

$$2NH_3 + 4O_2 \rightarrow N_2O_5 + 3H_2O$$

This involves the loss of eight electrons, $N(-III)$ $s^2p^6$, to $N(V)$, $s^0p^0$.

*Paragraph 4. The ammonium cation in its salts has been widely used as smelling salts in medicine and as a fertiliser in agriculture, both uses due to the available ammonia provided. Why the ammonia is so readily available is part of the unpredictable chemistry of the $NH_4^+$ cation:*

$$NH_4^+ + Heat \rightarrow NH_3 + H^+$$

*Conclusion: The $NH_4^+$ cation is an excellent example of a molecule, cation or anion to illustrate how 80% of its simple chemistry can be predicted from a knowledge of the electron configuration of the central nitrogen atom.*

## CONCLUSIONS

By using the *knowledge* inherent in the valence shell electron configuration of the elements of the Periodic Table, a great deal of this simple chemistry can be predicted. By using the logical sequence of the Working Method (Table 7.2), associated with the *Features of Interest* approach and presented in the Spider Diagrams of Figure 7.8 (a)–(h), the need to memorise a great deal of this factual chemistry is avoided and yet a significant amount of the chemistry of simple molecules, cations and anions can be attractively presented. The basic chemistry of a Spider Diagram can readily form the 'outline' for writing an essay or presenting a talk on any one of the compounds involved and conveys the impression that the student knows some chemistry.

## SUGGESTED WAYS FORWARD

### Phase II – Features of Interest

The *Features of Interest* approach, *via* the format of the Working Method/Spider Diagram, (Figure 7.1), offers a structured approach to the presentation of the chemistry of a simple molecule, cation or anion at a First Year Chemistry level. It encourages students to think about factual chemistry in a structured way from the start of their chemistry course, an approach that can then be developed in the following years of their course. The simple *Features of Interest* approach, Phase I of Figure 7.1, divides into two main sections, namely, stereochemistry and electronic properties and suggests a more advanced Phase II – *Features of Interest* Spider Diagram as shown in Figure 7.11. The stereochemistry of a molecule naturally leads to the point group symmetry, the modes of vibration, the NMR spectra and ultimately to structure determination. The electronic properties lead to theories of bonding, electronic spectra and magnetism. Each of the topics of Figure 7.11 can itself be developed as a special topic Spider Diagram, which at a second year level would still involve a relatively elementary treatment, but could be more detailed with succeeding years. Thus the structure determination section might only involve $X$-ray powder techniques in second year, space group determination in third year and single-crystal $X$-ray structure determination in the final year. The use of the Phase II Spider Diagram of Figure 7.11 is illustrated for the chemistry of $[Ni(NH_3)_6]SO_4$ in Figure 7.12.

Used in this way the phase I *Features of Interest* approach of the structure of a Spider Diagram introduced at the first year level can be readily extended into the succeeding years of an undergraduate chemistry course and provides a basis for instruction in essay writing or talk presentation.

## Phase III – Features of Interest

The Phase I and II Features of Interest approach to the chemistry of individual simple molecules, cations and anions (Table 7.1) can be used to describe the wider chemistry of the elements of the main group and transition metals, by presenting this chemistry in terms of a selection of simple compounds, each described by their individual Features of Interest approach developed in Phases I and II. Figure 7.13 suggests a phase III *Features of Interest* approach to the descriptive chemistry of a main group (or transition metal element), in which the representative simple compounds [Table 7.3(a) and (b)], are still 'connected' *via* the electron configuration of the central element in determining the oxidation numbers observed. The addition of a Spider Diagram of the reactions of the *element* provides not only methods of preparation of some simple compounds of the element, but illustrates the range of variable valence observed, in the compounds formed. An initial section on the occurrence of the element, its preparation, the electron configuration of the element, its variable valence and oxidation states, can introduce the Spider Diagram. A final section on the uses of the element in the laboratory and in industry completes the Spider Diagram.

Figure 7.14 illustrates the *Features of Interest* approach to the main group element carbon. Four simple compounds, $CH_4$, CO, $CO_2$ and $H_2CO_3$, are selected to illustrate the chemistry, two of which have already been given as examples of simple molecule Spider Diagrams in Figure 7.8(a) and (h); those for CO and $CO_2$ are illustrated in Figures 7.15 and 7.16. The chemistry of $CH_4$ illustrates the most reduced oxidation state of carbon, $-IV$, while $CO_2$ and $H_2CO_3$ illustrate the highest

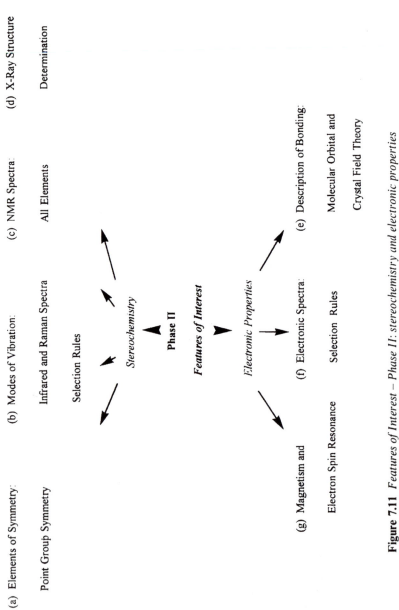

**Figure 7.11** *Features of Interest – Phase II: stereochemistry and electronic properties*

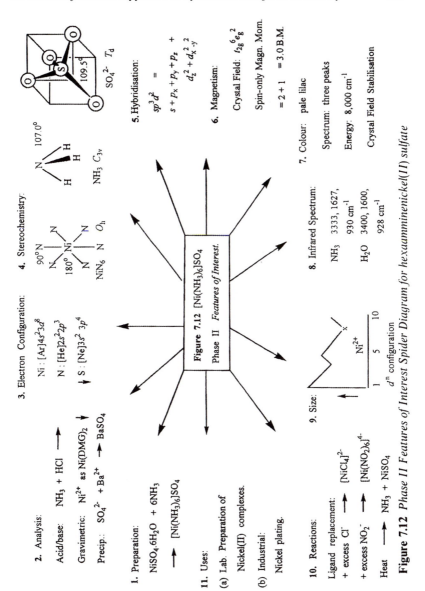

**2. Analysis:**

Acid/base: $NH_3 + HCl \longrightarrow$

Gravimetric: $Ni^{2+}$ as $Ni(DMG)_2 \longrightarrow$

Precip.: $SO_4^{2-} + Ba^{2+} \longrightarrow BaSO_4$

**3. Electron Configuration:**

Ni : $[Ar]4s^2 3d^8$

N : $[He]2s^2 2p^3$

S : $[Ne]3s^2 3p^4$

**4. Stereochemistry:**

$NH_3$ $C_{3v}$

$NiN_6$ $O_h$

**5. Hybridisation:**

$sp^3 d^2 =$

$s + p_x + p_y + p_z + d_z^2 + d_{x^2-y^2}^2$

$SO_4^{2-}$ $T_d$

**6. Magnetism:**

Crystal Field: $t_{2g}^6 e_g^2$

Spin-only Magn. Mom.

$= 2 + 1 = 3.0$ B.M.

**7. Colour:** pale lilac

Spectrum: three peaks

Energy: 8,000 cm$^{-1}$

Crystal Field Stabilisation

**8. Infrared Spectrum:**

$NH_3$ 3333, 1627, 930 cm$^{-1}$

$H_2O$ 3400, 1600, 928 cm$^{-1}$

**9. Size:**

**1. Preparation:**

$NiSO_4 \cdot 6H_2O + 6NH_3$

$\longrightarrow [Ni(NH_3)_6]SO_4$

**11. Uses:**

(a) Lab. Preparation of Nickel(II) complexes.

(b) Industrial: Nickel plating.

**10. Reactions:**

Ligand replacement:

$+$ excess $Cl^- \longrightarrow [NiCl_4]^{2-}$

$+$ excess $NO_2^- \longrightarrow [Ni(NO_2)_6]^{4-}$

Heat $\longrightarrow NH_3 + NiSO_4$

**Figure 7.12** [Ni(NH$_3$)$_6$]SO$_4$

Phase II *Features of Interest.*

**Figure 7.12** *Phase II Features of Interest Spider Diagram for hexaamminenickel(II) sulfate*

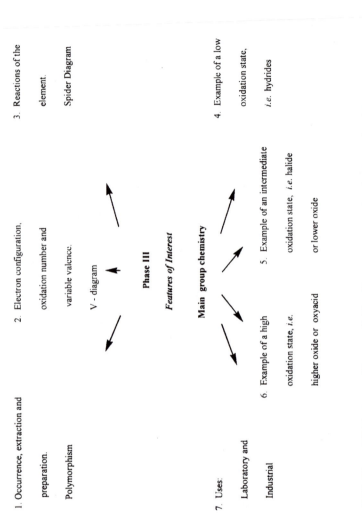

**Figure 7.13** *Phase III Features of Interest Spider Diagram – general main group chemistry*

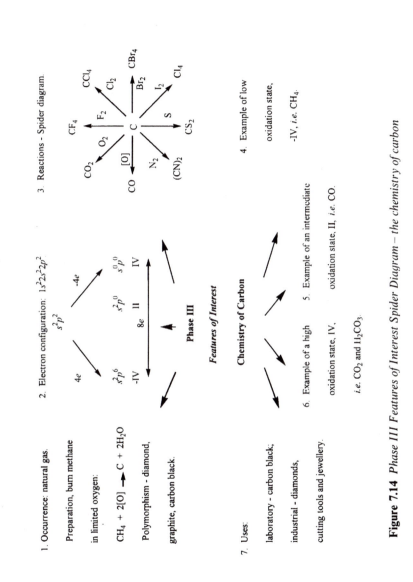

1. Occurrence: natural gas.

Preparation, burn methane

in limited oxygen:

$CH_4 + 2[O] \longrightarrow C + 2H_2O$

Polymorphism - diamond,

graphite, carbon black.

7. Uses:

laboratory - carbon black;

industrial - diamonds,

cutting tools and jewellery.

2. Electron configuration: $1s^2 2s^2 2p^2$

$s^2 p^2$

$-4e$

$4e$

$s^2 p^6$        $s^2 p^0$        $s^0 p^0$

$-IV$        II        IV

$8e$

**Phase III**

**Features of Interest**

**Chemistry of Carbon**

3. Reactions - Spider diagram.

CF_4    CCl_4    CBr_4

CO_2    O_2    F_2    Cl_2

[O]    C    Br_2    I_2

CO    N_2    S    Cl_4

(CN)_2    CS_2

4. Example of low

oxidation state,

-IV, *i.e.* $CH_4$.

5. Example of an intermediate

oxidation state, II, *i.e.* CO.

6. Example of a high

oxidation state, IV,

*i.e.* $CO_2$ and $H_2CO_3$.

**Figure 7.14** *Phase III Features of Interest Spider Diagram – the chemistry of carbon*

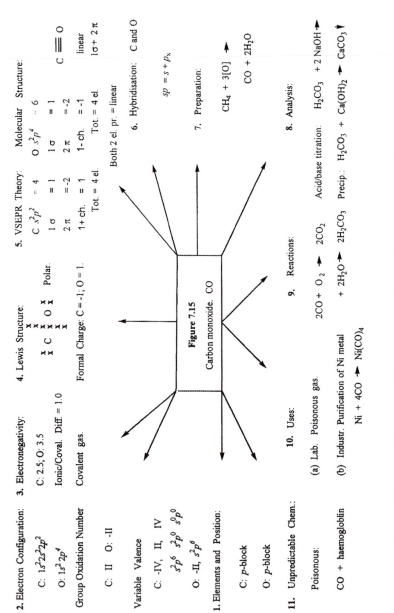

**Figure 7.15** *Phase III Features of Interest Spider Diagram for carbon monoxide*

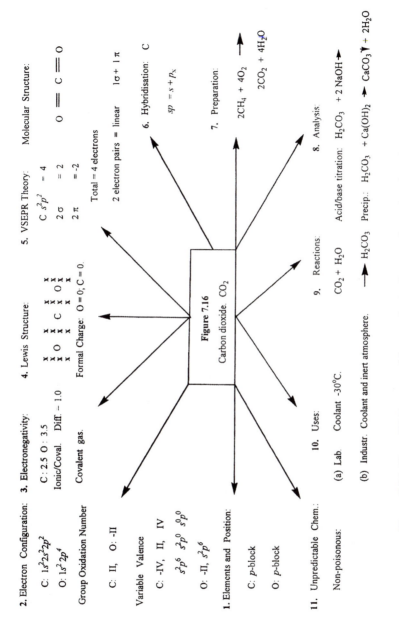

**2. Electron Configuration:**

C: $1s^2 2s^2 2p^2$

O: $1s^2 2p^4$

**Group Oxidation Number**

C: II, O: -II

**Variable Valence**

C: -IV, II, IV

$s^2 p^6$   $s^2 p^0$   $s^0 p^0$

O: -II, $s^2 p^6$

**1. Elements and Position:**

C: *p*-block

O: *p*-block

**3. Electronegativity:**

C: 2.5 O: 3.5

Ionic/Coval. Diff – 1.0

Covalent gas.

**4. Lewis Structure:**

$$O :: C :: O$$

Formal Charge: O = 0; C = 0.

**5. VSEPR Theory:**

$C \; s^2 p^2 \; = 4$

$2\sigma \qquad = 2$

$2\pi \qquad = -2$

Total = 4 electrons

2 electron pairs = linear    $1\sigma + 1\pi$

**Molecular Structure:**

$$O = C = O$$

**6. Hybridisation:** C

$sp = s + p_x$

**7. Preparation:**

$$2CH_4 + 4O_2$$

$$2CO_2 + 4H_2O$$

**8. Analysis:**

Acid/base titration: $H_2CO_3$ + 2 NaOH

Precip.: $H_2CO_3$ + Ca(OH)$_2$ → CaCO$_3$↓ + 2H$_2$O

**9. Reactions:**

$$CO_2 + H_2O \longrightarrow H_2CO_3$$

**10. Uses:**

(a) Lab.   Coolant -30°C.

(b) Industr. Coolant and inert atmosphere.

**11. Unpredictable Chem.:**

Non-poisonous:

**Figure 7.16**

Carbon dioxide. $CO_2$

**Figure 7.16** *Phase III Features of Interest Spider Diagram for carbon dioxide*

**Table 7.3** *Suggested simple compounds and complexes to illustrate a Features of Interest approach to (a) main group chemistry and (b) transition metal chemistry.*

(a) *Main Group Chemistry:*

| | | | | | | |
|---|---|---|---|---|---|---|
| Na | $s^1$ | Na | NaH | $Na_2O$ | NaOH | NaCl | $Na_2SO_4$. |
| Be | $s^2$ | Be | $BeH_2$ | BeO | $Be(OH)_2$ | $BeCl_2$ | $BeSO_4$ |
| Mg | $s^2$ | Mg | $MgH_2$ | MgO | $Mg(OH)_2$ | $MgCl_2$ | $MgSO_4$ |
| B | $s^2p^1$ | B | $BH_3$ | $B_2O_3$ | $H_3BO_3$ | $BCl_3$ | $B_2(SO_4)_3$ |
| Al | $s^2p^1$ | Al | $AlH_3$ | $Al_2O_3$ | $Al(OH)_3$ | $AlCl_3$ | $Al_2(SO_4)_3$ |
| C | $s^2p^2$ | C | $CH_4$ | CO | $CO_2$ | $CCl_4$ | $H_2CO_3$ |
| Si | $s^2p^2$ | Si | $SiH_4$ | SiO | $SiO_2$ | $SiCl_4$ | $H_4SiO_4$ |
| N | $s^2p^3$ | $N_2$ | $NH_3$ | $N_2O_3$ | $N_2O_5$ | $NO_2^-$ | $HNO_3$ |
| P | $s^2p^3$ | $P_4$ | $PH_3$ | $P_2O_3$ | $P_2O_5$ | $PO_4^{3-}$ | $H_3PO_4$ |
| S | $s^2p^4$ | $S_8$ | $SH_2$ | $SO_2$ | $SO_3$ | $SF_6$ | $H_2SO_4$ |
| Cl | $s^2p^5$ | $Cl_2$ | ClH | $Cl_2O_6$ | $Cl_2O_7$ | $HClO_3$ | $HClO_4$ |

(b) *Transition Metal Chemistry:*

| | | | | | | |
|---|---|---|---|---|---|---|
| Cr | $s^2d^4$ | Cr | $Cr_2O_3$ | $CrO_3$ | $CrCl_3$ | $K_2Cr_2O_7$ | $[Cr(OH_2)_6]^{3+}$ |
| Mn | $s^2d^5$ | Mn | MnO | $MnO_2$ | $MnCl_2$ | $KMnO_4$ | $[Mn(OH_2)_6]^{2+}$ |
| Fe | $s^2d^6$ | Fe | FeO | $Fe_3O_4$ | $FeCl_2$ | $FeCl_3$ | $[Fe(OH_2)_6]^{3+}$ |
| Ni | $s^2d^8$ | Ni | $NiO_2$ | $NiO_2$ | $NiCl_2$ | $[Ni(OH_2)_6]^{2+}$ | $K_2[NiCl_4]$ |
| Cu | $s^2d^9$ | Cu | $Cu_2O$ | CuO | CuCl | $CuCl_2$ | $[Cu(OH_2)_6]^{2+}$ |

oxidation state, IV, connected by the 8e electron difference of $s^2p^6\ s^0p^0$. The chemistry of CO illustrates the intermediate II oxidation state stabilised by the *pseudo* inert pair effect, $s^2p^0$. The Spider Diagram of the reactions of carbon illustrates the chemistry of the element carbon and the range of variable valence observed. An introduction describes the occurrence of the element, its extraction and preparation, and the $s^2p^2$ electron configuration allows the prediction of its variable valence in a V-diagram (Figure 4.1). The addition of a uses section completes an overview of the chemistry of carbon.

Figure 7.17 illustrates the general *Features of Interest* approach to the chemistry of a transition metal element, while Figure 7.18 applies the approach to the chemistry of iron, involving two simple compounds, $Fe^{II}Cl_2$ and $Fe^{III}Cl_3$ and two simple complexes, $[Fe^{II}(OH_2)_6]SO_4$ and $[Fe^{III}(OH_2)_6](NO_3)_3$ [Table 7.3(b)]. The phase I *Features of Interest* Spider Diagrams of the iron(II) compounds have already been illustrated in Figures 7.8(f) and (g), and the corresponding iron(III) diagrams are illustrated in Figures 7.19 and 7.20. Only two common oxidation states occur with iron, II and III, but the less common higher oxidation state of IV arises in the reaction of the element with the most electronegative element fluorine and is illustrated in the reaction Spider Diagram of Figure 7.18. The addition of an occurrence, extraction and preparation section of iron, plus the electron configuration and and variable valence of iron, opens the Spider Diagram. A uses section completes the Spider Diagram of the overview of the chemistry of iron.

### The Advantages of the *Features of Interest* Approach

- It brings together the *connectivities* approach to predicting the chemistry of *simple molecules*, cations and anions.
- The structured approach of the Working Method is inherent in the Spider Diagram, phase I, which can be extended to more advanced species, Phase II and particularly, phase III, to the factual descriptive main group or transition metal chemistry of the elements.
- The structure of the *Features of Interest* Spider Diagram approach may be used to form the basis of a chemical essay, report or talk and provides students with the confidence that they know some chemistry and helps to reduce the need to memorise factual chemistry to a more manageable amount.

### The Limitations of the *Features of Interest* Approach

- The formal structure of the *Features of Interest* Spider Diagram suggests a rather rigid approach to learning chemistry, but once the students understand the approach inherent in the connectivities of the Spider Diagram, they are encouraged to construct their own versions of the Working Method/Spider Diagram.
- The ability to *predict* the simple chemistry of a molecule, cation or anion implies that all the properties of these species are predictable, which is clearly *not* the case. Only *ca.* 80% predictability is suggested, even at this elementary level of first year chemistry and even this percentage falls in the higher years of an undergraduate course. However, the availability of the *Features of Interest* Working Method/Spider Diagram structure enables the student to get started and show that he/she knows some chemistry.

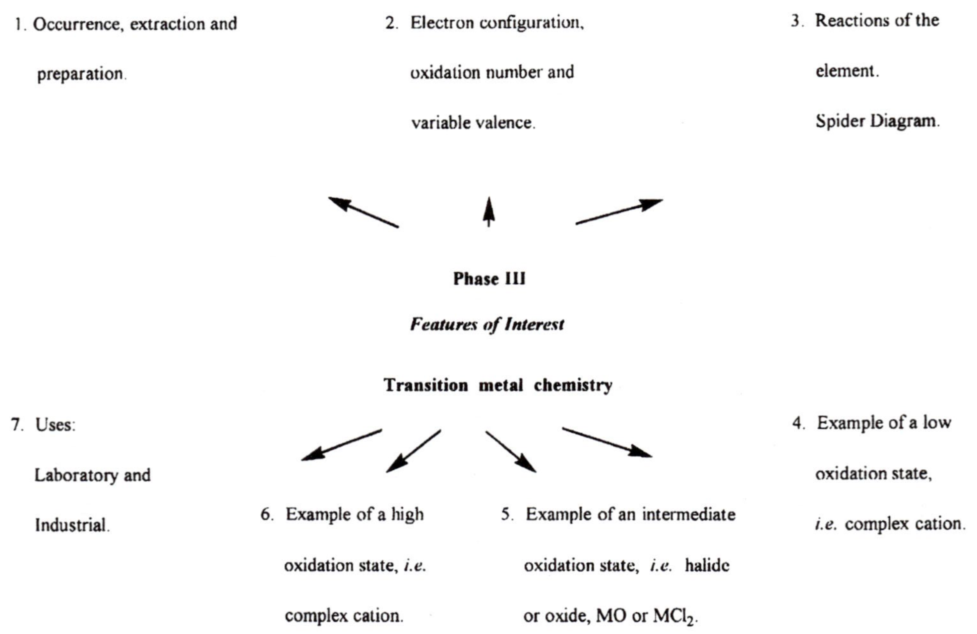

**Figure 7.17** *Phase III Features of Interest Spider Diagram – general transition metal chemistry*

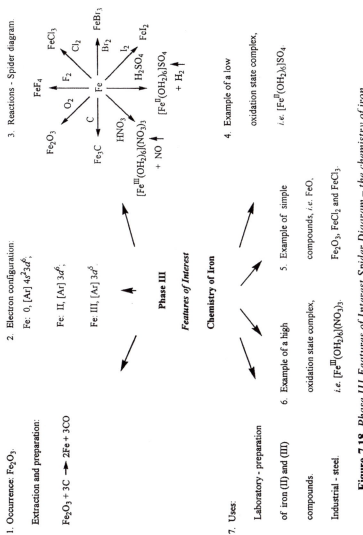

**Figure 7.18** *Phase III Features of Interest Spider Diagram – the chemistry of iron*

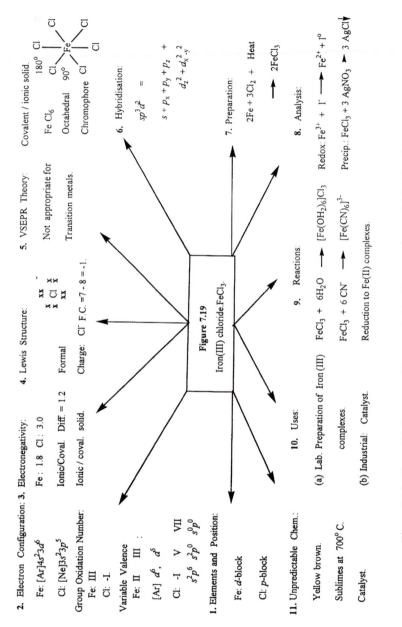

**Figure 7.19** *Phase III Features of Interest Spider Diagram for iron(III) chloride*

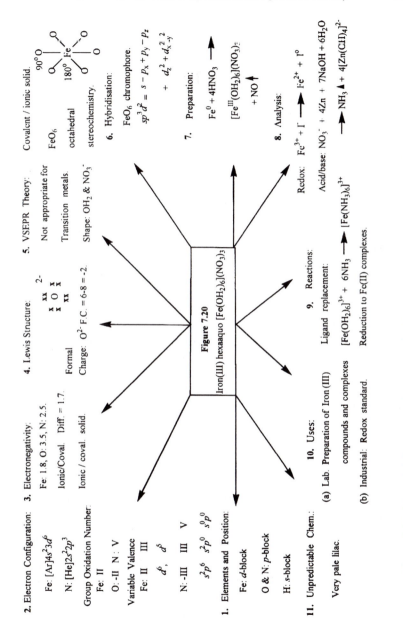

**2. Electron Configuration:**

Fe: [Ar]$4s^2 3d^6$

N: [He]$2s^2 2p^3$

**Group Oxidation Number:**

Fe: II

O: -II   N: V

**Variable Valence**

Fe: II   III

$d^6$, $d^5$

N: -III   III   V

$s^2 p^6$   $s^2 p^0$   $s^0 p^0$

**1. Elements and Position:**

Fe: *d*-block

O & N: *p*-block

H: *s*-block

**11. Unpredictable Chem.:**

Very pale lilac.

**3. Electronegativity:**

Fe: 1.8, O: 3.5, N: 2.5.

Ionic/Coval.   Diff = 1.7.

Ionic / coval. solid.

**4. Lewis Structure:**

Formal

Charge: $O^{2-}$ F.C. = 6-8 = -2.

**5. VSEPR Theory:**

Not appropriate for Transition metals.

Shape: $OH_2$ & $NO_3^-$

Covalent / ionic solid.

$FeO_6$

octahedral

stereochemistry.

$90^\circ$ $180^\circ$ Fe

**6. Hybridisation:**

$FeO_6$ chromophore.

$$sp^3 d^2 = s - p_x + p_y - p_z + d_{z^2} + d_{x^2-y}^2$$

**7. Preparation:**

$$Fe^0 + 4HNO_3 \longrightarrow$$

$$[Fe^{III}(OH_2)_6](NO_3)_3$$

$$+ NO \uparrow$$

**8. Analysis:**

Redox: $Fe^{3+} + I^- \longrightarrow Fe^{2+} + I^0$

Acid/base: $NO_3^- + 4Zn + 7NaOH + 6H_2O \longrightarrow NH_3 \uparrow + 4[Zn(ClI)_4]^{2-}$

**9. Reactions:**

Ligand replacement:

$$[Fe(OH_2)_6]^{3+} + 6NH_3 \longrightarrow [Fe(NH_3)_6]^{3+}$$

Reduction to Fe(II) complexes.

**10. Uses:**

(a) Lab. Preparation of Iron (III) compounds and complexes

(b) Industrial: Redox standard.

**Figure 7.20**

Iron(III) hexaaquo $[Fe(OH_2)_6](NO_3)_3$

**Figure 7.20** *Phase III Features of Interest Spider Diagram for hexaaquo iron(III) trinitrate*

# Appendices

**Appendix I** *References*

(a) *Texts and papers*:

1. H.F. Holtzclaw, W.R. Robinson and J.D. Odom, 'General Chemistry', D.C. Heath and Co., 1991.

2. J.E. Brady and J.R. Holum, 'Chemistry', John Wiley, 1993.

3. G.M. Bodner, L.H. Rickard and J.N. Spencer, 'Core Text Chemistry' Wiley, 1996.

4. E.C. Constable and A.E. Houseman, 'Chemistry', Addison, Wesley, Longman, 1997.

5. B. Murphy, C. Murphy and B.J. Hathaway, 'A Working Method Approach for Introductory Physical Chemistry Calculations', The Royal Society of Chemistry, 1997.

6. B. Murphy, C. Murphy and B. Hathaway, 'Can Computer Aided Learning Benefit the Teaching of Chemistry', Software Reviews, 1997, **15**, 12–16.

(b) *CAL Tutorials*:

Listing of the ten CAL tutorials available on Connectivities in Inorganic Chemistry.

1. Mole: Materials – States of Matter – Elements and Compounds – Law of Constant Composition – Law of Conservation of Mass. Mole – Avogadro's Number – Empirical and Molecular Formula. Molarity in Solution.

2. VolCal1: Types of Reactions – Acid/Base – Precipitation – Redox. Given Stoichiometry – Volumetric Calculations – Molarity/Volume/Weight.

3. AtomStru: Structure of the Atom – Dalton, Thompson and Rutherford Models – Atomic Number – Atomic Weights – Isotopes.

4. BohrPT: Bohr Model of the Atom – Four Quantum Numbers – Electron Configuration of the Atom – Electron Shells – Shapes of Orbitals – Wave Nature of the Electron – Wave Functions, Radial and

Angular Parts. Many Electron atoms – Hund's Rules – Build-up Process – Maximum Multiplicity – Periodic Table – *s*-, *p*- and *d*-block Elements – Long Form of the Periodic Table.

5. PhysProp: Effective Nuclear Charge – Ionisation energy – Electron Affinity – Covalent and Ionic Radii – Electronegativity – Orbital Energies and Promotional energies.

6. ChemProp: Group Oxidation Number – Inert Gas Core – 8-electron configuration – V-type Diagrams – Group Oxidation State – Variable Valence. Ligands – $[M(OH_2)_6]^{2+}$ – Oxyanions.

7. Lewis: Covalent and Ionic Bonds – Lewis Structures – Octet Rule – Cations and Anions – Lone Pairs – Incomplete Octets – Expanded Octets – Double and Triple Bonds – Oxyacids – Resonance.

8. Shape: Valence State Electron Pair Repulsion Theory – Application to Molecules, Anions and Cations of Main Group Elements – Molecules with Lone Pairs – Oxyanions. Hybridisation -$sp$, $sp^2$, $sp^3$, $sp^3d$ and $sp^3d^2$ Hybrid Orbitals – Multiple Choice Questions.

9. Features of Interest: Predicting the descriptive chemistry of simple molecules, cations and anions. Working Method, Features of Interest Spider Diagram, example $NH_4^+$ cation. Writing a Chemical Essay, $NH_4^+$ example. Shorter examples: $SO_4^{2-}$ anion, $PCl_5$ and $[Mn(OH_2)_6]^{2+}$ cation.

10. VolCal2: Limiting Reactions – Calculation of Stoichiometry Factors – Working Method – Volumetric Calculations.

These ten tutorials are available free of charge on the internet at:
   URL:http://nitec.dcu.ie/~chemlc/CAL2.html
The eight physical chemistry tutorials, ref. 5, are also available free at this address and have been reviewed in CTI Chemistry *Software Reviews*, 1997, **16**, 4-15.

**Appendix II** *Suggested questions for tutorials, discussion classes or examinations*:

(a) *Discussion class or tutorial material*:

   (1) Balancing equations:
In the reaction, $v_A \cdot A + v_B \cdot B \rightarrow$, what is the ratio of the stoichiometry factors $v_A : v_B$ ?
A. Acid/base: for the following acids (A) and bases (B) determine the $v_A : v_B$ ratio

| A | HCl | $H_2SO_4$ | $H_3PO_4$ |
|---|---|---|---|
| B NaOH | 1:1 | 1:2 | 1:3 |
| Ba(OH)$_2$ | 2:1 | 1:1 | 2:3 |
| Al(OH)$_3$ | 3:1 | 3:2 | 1:1 |
| Na[Al(OH)$_4$] | 4:1 | 2:1 | 4:3 |
| Ba[Al(OH)$_5$] | 5:1 | 5:2 | 5:3 |

B. Precipitation: for the following reactants A with B, determine the $v_A:v_B$ ratio.

| A | $AgNO_3$ | $Ag_2SO_4$ | $Ag_3PO_4$ |
|---|---|---|---|
| B NaCl | 1:1 | 1:2 | 1:3 |
| BaCl$_2$ | 2:1 | 1:1 | 2:3 |
| AlCl$_3$ | 3:1 | 3:2 | 1:1 |
| PCl$_4^+$ | 4:1 | 2:1 | 4:3 |
| PCl$_5$ | 5:1 | 5:2 | 5:3 |
| PCl$_6^-$ | 6:1 | 3:1 | 2:1 |

Alternative questions can be generated using the $Br^-$ and $I^-$ anions.

C. Redox reactions: for the following reactants A with B, determine the $v_A:v_B$ ratio.

| A | $KMnO_4$ | $K_2Cr_2O_7$ |
|---|---|---|
| B Fe$^{II}$ | 1:5 | 1:6 |
| SO$_3^{2-}$ | 2:5 | 2:6 |
| FeSO$_3$ | 3:5 | 3:6 |
| Fe$_2$(SO$_3$)$_3$ | 6:5 | 1:1 |
| K$_3$[Fe(SO$_3$)$_3$] | 6:5 | 1:1 |
| K$_4$[Fe(SO$_3$)$_3$] | 7:5 | 7:6 |

Alternative questions can be generated using the $NO_2^-$ or $ClO_3^-$ anions.

In handouts to students the stoichiometry ratios are omitted.

(2) Multiple choice questions:
What are the stoichiometry factors, $v_A:v_B$, for the reaction: $v_A \cdot A + v_B \cdot B \rightarrow$, of $K_3[CuCl_5]$ and $Ag_3PO_4$ ?
  [A] 2:3    [B] 3:5    [C] 5:3    [D] 3:3

(3) Further suggested simple compounds, cations or anions:

Molecules: $CH_4$; $NH_3$; $OH_2$; $BF_3$; $PCl_5$; $PCl_3$; $SF_6$; $BeCl_2$.
        $SO_2$; $SO_3$; $H_2CO_3$; $HNO_2$; $HNO_3$; $H_3PO_4$; $H_2SO_4$;
        $H_2SO_3$; $HClO_4$; $HClO_3$.

Cations:    $NH_4^+$; $PCl_4^+$; $[Mn(OH_2)_6]^{2+}$; $[Fe(OH_2)_6]^{2+}$.

Anions:     $BF_4^-$; $PCl_6^-$; $CO_3^{2-}$; $NO_3^-$; $NO_2^-$; $SO_4^{2-}$; $SO_3^{2-}$; $PO_4^{3-}$; $PO_3^{3-}$; $ClO_3^-$; $ClO_4^-$.

(b) *Examination questions*:

(1) For each of the elements indicated by the asterisk (*) listed below:
$P^*Cl_5$; $N^*H_3$; $C^*O_3^{2-}$; $[Mn^*(OH_2)_6]^{2+}$.

(a) Give the electron configuration of the free element and of the element in the given compound, and the oxidation number of the element in the compound.

(b) Draw the Lewis structure.

(c) Predict the stereochemistry. Justify your prediction with VSEPR theory or otherwise.

(d) Indicate the hybridisation appropriate to each stereochemistry. Specify the atomic orbitals involved and sketch the hybrid orbitals.

(e) Suggest a method of preparation and one method of analysis for each compound, cation or anion.

(2) Draw a Spider Diagram to develop the simple chemistry of two of the following: $SF_6$; $NH_3$; $H_2SO_4$; $[Fe(OH_2)_6]^{3+}$.

(3) Draw a Spider Diagram to develop the simple chemistry of (a) nitrogen and (b) copper.

# Periodic Table of the Elements

Reproduced with the kind permission of Glaxo Wellcome plc.

| Group | 1 | 2 | 3 | 4 | 5 | 6 | 7 | 8 | 9 | 10 | 11 | 12 | 13 | 14 | 15 | 16 | 17 | 18 |
|---|---|---|---|---|---|---|---|---|---|---|---|---|---|---|---|---|---|---|
| **1s** | 1 H 1·0079 | | | | | | | | | | | | | | | | | 2 He 4·0026 |
| **2s / 2p** | 3 Li 6·941 | 4 Be 9·01218 | | | | | | | | | | | 5 B 10·81 | 6 C 12·011 | 7 N 14·0067 | 8 O 15·9994 | 9 F 18·9984 | 10 Ne 20·179 |
| **3s / 3p** | 11 Na 22·98977 | 12 Mg 24·305 | | | | | | | | | | | 13 Al 26·9815 | 14 Si 28·0855 | 15 P 30·9738 | 16 S 32·06 | 17 Cl 35·453 | 18 Ar 35·948 |
| **4s / 4p** | 19 K 39·0983 | 20 Ca 40·08 | 21 Sc 44·9559 | 22 Ti 47·88 | 23 V 50·9415 | 24 Cr 51·996 | 25 Mn 54·938 | 26 Fe 55·847 | 27 Co 58·9332 | 28 Ni 58·69 | 29 Cu 63·546 | 30 Zn 65·38 | 31 Ga 69·72 | 32 Ge 72·59 | 33 As 74·9216 | 34 Se 78·96 | 35 Br 79·904 | 36 Kr 83·80 |
| **5s / 5p** | 37 Rb 85·4678 | 38 Sr 87·62 | 39 Y 88·9059 | 40 Zr 91·22 | 41 Nb 92·9064 | 42 Mo 95·94 | 43 Tc (98) | 44 Ru 101·07 | 45 Rh 102·9055 | 46 Pd 106·42 | 47 Ag 107·868 | 48 Cd 112·41 | 49 In 114·82 | 50 Sn 118·69 | 51 Sb 121·75 | 52 Te 127·60 | 53 I 126·9045 | 54 Xe 131·29 |
| **6s / 6p** | 55 Cs 132·9054 | 56 Ba 137·33 | 57 'La 138·9055 | 72 Hf 178·49 | 73 Ta 180·9479 | 74 W 183·85 | 75 Re 186·207 | 76 Os 190·2 | 77 Ir 192·22 | 78 Pt 195·08 | 79 Au 196·9665 | 80 Hg 200·59 | 81 Tl 204·383 | 82 Pb 207·2 | 83 Bi 208·9804 | 84 Po (209) | 85 At (210) | 86 Rn (222) |
| **7s** | 87 Fr (223) | 88 Ra 226·0254 | 89 'Ac 227·0278 | | | | | | | | | | | | | | | |

( ) mass numbers of most stable isotope

**' LANTHANUM SERIES**

| 4f | 58 Ce 140·12 | 59 Pr 140·9077 | 60 Nd 144·24 | 61 Pm (145) | 62 Sm 150·36 | 63 Eu 151·96 | 64 Gd 157·96 | 65 Tb 158·9254 | 66 Dy 162·50 | 67 Ho 164·9304 | 68 Er 167·26 | 69 Tm 168·9342 | 70 Yb 173·04 | 71 Lu 174·967 |
|---|---|---|---|---|---|---|---|---|---|---|---|---|---|---|

**' ACTINIUM SERIES**

| 5f | 90 Th 232·0391 | 91 Pa 231·0359 | 92 U 238·0369 | 93 Np 237·0482 | 94 Pu (244) | 95 Am (243) | 96 Cm (247) | 97 Bk (247) | 98 Cf (251) | 99 Es (252) | 100 Fm (257) | 101 Md (258) | 102 No (259) | 103 Lr (260) |
|---|---|---|---|---|---|---|---|---|---|---|---|---|---|---|

# Subject Index